University Responsibility for the Adjudication of Research Misconduct

Stefan Franzen

University Responsibility for the Adjudication of Research Misconduct

The Science Bubble

Springer

Stefan Franzen
Department of Chemistry
North Carolina State University
Raleigh, NC, USA

ISBN 978-3-030-68062-6 ISBN 978-3-030-68063-3 (eBook)
https://doi.org/10.1007/978-3-030-68063-3

This Springer imprint is published by the registered company Springer Nature Switzerland AG
The registered company address is: Gewerbestrasse 11, 6330 Cham, Switzerland

Foreword

Several years ago, we made a terrific discovery in our university laboratory – not a world-changing discovery, not a cure for cancer or a new way to harness the energy of the sun, but a bright spot in our small corner of the universe of science that would have formed the basis of an attractive publication. After our initial exuberance subsided, and after discussing how to organize our findings, a conscientious doctoral student recognized that our "discovery" was merely an artifact of a miscalibrated instrument, a consequence of a "+1" in a computer program that should have been a "−1." The student entered my office cautiously to tell me what he had determined, concerned as the bearer of bad news; he thought that I would be disappointed by the apparently unfortunate turn of events. Instead, I was thrilled that a student in my charge deliberately tried to rain on our parade. There was no better evidence that I was doing my job, a judgment that was made plain by his extra detective work, a double check of all systems to make sure that our presumed good fortune really was good fortune. It was not.

The process of experimental science includes a design strategy, the collection of data, the interpretation of the results, and last but not least, an aggressive effort to roll back any uplifting news that should follow from the first three steps. An uncritical presumption that good fortune really is good fortune is not proper conduct for a professional scientist. It is adolescent self-deception. That is why the fourth step, an effort to reject a noteworthy result for credible reasons, is essential. There are few professions that operate in this witheringly self-critical manner. A banker does not tell his or her client the reasons why a recommended investment may be a bad risk. A defense attorney does not argue in front of a jury that a new piece of evidence may implicate rather than exonerate a client. In other fields, we expect some actors to sell, not discredit, good news. This distinction is what makes Science special. This is why scientists have historically earned great public confidence. Scientists expect that other scientists tell the truth, as best as they are able, all the time, no matter what. Nevertheless, cynicism may be growing in the public mind as reports of corporate intrusions into research, among other conflicts of interest, are increasingly analyzed in the media. Scientists should be concerned about the erosion of public support, since taxpayers are unwittingly the recipients of many of the bills that scientists accumulate in expensive university and government laboratories.

Sometimes, our capacity as scientists for self-criticism reaches its limit and we deceive ourselves. Scientists are human and wishful self-deception is a human quality. Discoveries and apparent discoveries are like drugs because they create feelings of exuberance. We want these feelings to last forever. Therefore, it is essential that in today's highly collaborative scientific society there are critical participants or witnesses who work against what our brain's pleasure centers crave, good news. At some point, every scientist needs someone else's help to sober up. We stay vigilant for one another and, in fact, we spend many hours a week in this kind of work, *gratis,* through the peer review of grant proposals to federal funding agencies and manuscripts submitted to academic journals for publication, and in meetings with students and colleagues about the progress of research projects. However, when an error of yours has been revealed, the singular way forward is to first stop, absorb the impact of the mistake, tell it, and fix it. Failure to rigorously implement these steps marks the moment when self-deception may shade into misconduct. The misguided scientist may prefer to keep selling the original story because mistakes are embarrassing and the consequences of being wrong may be personally and professionally inconvenient. No one *likes* to admit being wrong.

What if the happy science story you are selling, as in persuading, is also *selling*, being exchanged for money? What if patents and intellectual property are at stake if your story can no longer be supported in light of new evidence? How much harder will it become to do the right thing? A lot. If telling the truth will cost money, scientists may try to wrap unambiguous mistakes in the cloth of equivocality. The expectation that scientists are truthful has been, in part, a consequence of the fact that science at universities has been long free from commercial interests. But this changed. Financial entanglements of companies and intellectual property are now the norms at research universities and these entanglements have coincided with a disturbing rise in scientific misconduct [1]. The correlation between academic finance and academic fraud is sufficiently strong that Stefan Franzen asks in *The Science Bubble* whether science can survive the continued intrusion of commercial interests in academic work.

If forced to mark when the modern American university became beholden to outside financial interests, many would cite the passage of the 1980 Bayh-Dole Act, legislation sponsored by United States Senators Birch Bayh (D-Indiana) and Bob Dole of (R-Kansas) that was intended to bolster American competitiveness in an increasingly globalized economy. Prior to Bayh and Dole, universities and their employees could not directly profit from discoveries based on federally funded research that turned out to have a value in the marketplace. After all, if the research was purchased by taxpayers at universities that have already been exempted from taxes, selling discoveries back to the public while drawing profits for professors and their employers might seem to some like charging for the same thing in three different ways. In former times, it was presumed that benefits of publicly funded science research should serve the public welfare and that these benefits should freely accrue to those who supported the work, taxpayers. However, Congress recognized that practical benefits derived from basic research might never come to fruition unless the commercialization of technologies was incentivized. Creating a saleable technology requires more than a good idea somehow aligned with industrial interests. Making a product is hard, creative work. Bayh and Dole therefore sought to unleash the profit motive inside American colleges and universities. The promise of royalty riches for faculty as well as institutions of higher education, they surmised, would draw out insular professors pursuing curiosity-driven lines of inquiry with little concern for practical application, encouraging researchers to drive abstract discoveries to market, a job formerly conceded to industry. Bayh and Dole baited the hook. Researchers as well as university administrations bit hard. Some have argued that the commercialization of the university was well underway before Congress became involved [2], but Bayh-Dole certainly gave a legal justification and congressional imprimatur to those keen on a transformation of the social organization of American science.

Now that the train has left the station, so to speak, every research university has joined an "intellectual property arms race," patenting all manner of discoveries while hoping for purchase or license by a major corporation. This does happen, but so infrequently it is fair to characterize such activities as sanctioned gambling that is costly for institutions, and unconscionable for public institutions, in my view. High-salaried lawyers are required; patenting fees are steep. Most inventions, whether aggressively marketed or not, produce nothing in return. Nevertheless, a rising university administrator recently announced to her faculty that her chief ambition was to put "commercialization and entrepreneurship on steroids" [3]. The journalist Jennifer Washburn called sentiments of this kind part of a "foul wind" blowing across campuses [4]. In former generations, such a statement would have sounded discordant, even anti-academic. Today, it is met with applause.

Athletes take steroids because they are shortcuts to victory. Unfortunately, the creation of science cannot survive shortcuts. For science to work as it should – and we have some 350 years of science history to convince ourselves that it has been working very effectively – scientists must be as truthful as they are able. This is the *sine qua non* of science, but it is a stiffer standard if the business of science becomes entangled with business. As Ravetz presciently forecast in 1971, "the goals of a successful career in science can change from being a series of successful research projects made possible by a parallel series of adequate contracts, to being a series of successful research contracts made possible by a parallel series of adequate projects." Not only can this change happen, it has happened. University leaders, those who should be arguing for the preservation of academic values of honesty, transparency, and common welfare, are driving this rush to the marketplace. Laws simultaneously encourage universities to commercialize technology from academic research while simultaneously charging those institutions with the responsibility for monitoring the ethics of their own faculty. How can a university be objective if ethical issues arise in its own financial portfolio?

What has been the consequence of commercialization on steroids? When public universities decided to enter the common marketplace, state support for higher education began to plummet. By 2050, most states will likely be out of the business of funding universities altogether. The great American public universities may become nothing more than profit centers on state ledgers.

Author Stefan Franzen has experienced the consequences of the conflicts of interest at a contemporary research university at first hand. He coined the term "science bubble" to make an analogy with the real estate bubble that led to the worldwide financial crisis of 2008–2009. Junk science, invented to earn big grants, venture capital, and royalties, is akin junk mortgages. It has the capacity to poison the enterprise of science that is centuries old and international in scope. Junk science has been engineered by an academic culture that has adapted "perverse incentives" that are not only dollar-centric but based on superficial metrics of output rather than the richness of intellectual and educational outcomes [5]. For example, when Franzen and I were students, publications were sent aloft, so to speak, into the breeze; we were unsure of how or where they would land. Perhaps someone read them. Perhaps we would discover their fate/impact in the fullness of time. They were gifts to a community – even if they were gifts that nobody may have wanted. Whatever could be expected in return would be, at best, gratifications much delayed. Today, publications instantly begin to accumulate scores (#s of citations) on Google Scholar, checked in a moment. Gifts have been transformed into commodities – if not junk, publications have become junkifed. Selling junk science for money and creating science for the purpose of performance metrics has the potential to devalue all science. The science bubble will burst when it becomes widely recognized that scientific assets are grossly overvalued, and the public will no longer consider the support of science a virtue. Should we lose confidence in science when the future of humanity depends on the forecasts of scientists – climate change scientists and epidemiologists today are the most obvious examples – we do so at our peril.

One might presume that "grown-ups" are somewhere in charge, predisposed to prevent catastrophe. However, we all discovered during the financial crisis that regulators in the ratings industry worked hand in hand with the most corrupt bankers thereby making a bad situation worse. The tragi-comic narratives of many scientific misconduct investigations reveal that in science regulation, grown-ups are likewise few and far between. The Director of the Office of Research Integrity (ORI) for the Department of Health and Human Services (HHS), charged with integrity of ca. $30 billion in annual research expenditures, resigned in disgust, describing his office as "remarkably dysfunctional" and hamstrung by "secretive, autocratic, and unaccountable" supervisors [6].

The most consequential case of scientific misconduct or alleged scientific misconduct, a case that birthed the ORI in its present incarnation, is known as The Baltimore Affair. This name does not refer to site of the malfeasance, but rather David Baltimore, a prodigy of molecular biology celebrated for discoveries that will last for ages, who moreover served as the president of *two* leading American research universities. Baltimore collaborated with another senior scientist who was accused of misrepresentation of research data by a young coworker, Margot O'Toole. At the time that the allegation was made there was no formal procedure for reporting improper research practices; there was no "research misconduct system" such as it is called today. Frequently, the associated universities held perfunctory, *ad hoc* inquiries and declared that the rules had been obeyed. However, in the Baltimore case, evidence was strong enough that others became involved including the ORI and eventually even the Secret Service. The case was brought to the attention of a powerful critic, Michigan's John Dingell (D), the all-time longest-serving member of the U.S. House of Representatives. Battle lines were drawn, explosive hearings were held on the floor of Congress pitting two giants in their respective fields, forensic investigations were launched, the Department of Justice was asked to bring criminal charges (it didn't), and lives were upturned, none more so than O'Toole, the researcher who first came forward with concerns. For speaking out, she was driven from academic science altogether.

Howard Temin, a molecular biologist who shared the Nobel Prize with Baltimore (and Renato Delbucco), was asked by the historian Horace Freeland Judson to comment on the Baltimore Affair after the dust had settled. Temin said of the Baltimore failure:

> "When an experiment is challenged no matter who it is challenged by, it's your responsibility to check. That is an ironclad rule of science that, when you publish something, you are responsible for it. And one of the great strengths of American science...is that even the most senior professor if challenged by the lowliest technician or graduate student, is required to treat them seriously and to consider their criticisms." [7]

Baltimore did not check. If any member of a research team has a credible concern regarding a piece of research, then every member of the team must stop, take stock, and work either to affirm or refute the concerns raised. Scientists must not only tell the truth; they must be aggressive about testing the veracity of what they report and about what others report.

The Baltimore mistakes were nevertheless equivocal. The misdeeds of history's greatest science frauds have been incandescent by comparison [8, 9]. Why then was The Baltimore Affair "most consequential"? As Franzen describes herein, in the aftermath of Congressional hearings, federal science and regulatory agencies recognized that rules and procedures for adjudicating allegations of scientific misconduct were piecemeal and inadequate. The persecutions and exonerations of Baltimore highlighted a federal system that lacked guidelines and systematic procedures. The case was conducted over a 10-year period that coincided with writing of research misconduct regulations by various federal agencies under the guidance of the Office of Science and Technology Policy. The rules that were conceived and refined in the 1990s were compromises between government authorities keen on oversight, and scientists who feared the intrusion of ill-informed science police mucking up research. Confidentiality was the foremost consideration of those who were invested in protecting the interests of scientists fearful of false accusations. Secrecy, the anathema of good science, rules all those responsible for investigating improper science at universities and at funding agencies. Federal Offices of Inspectors General have conducted multiyear investigations without ever questioning those who brought concerns about complex matters to their attention in the first place. Regulators are required to operate in shadow, even if in ignorance, anonymizing public records that often can only be obtained – and only be obtained at *public* universities subject to state open records laws – by those with the means to sue for them. Investigators can never be questioned because there is no occasion for conversation. Communication is fractured in time and unidirectional.

Worse still, regulations for dealing with scientific misconduct investigations that developed in the wake of the Baltimore case presume that universities are intrinsically disinterested parties in such inquiries. This is absurd in the post Bayh-Dole world with universities trying to strike it rich through the exploitation of commercialized science. Regulations have encouraged universities to invent any process they should choose to adjudicate misconduct, processes that will naturally protect the considerable financial interests of the very institutions who created the rules of the game. If a university is monetarily invested in the outcome of a misconduct inquiry, it can declare that all has been fully and fairly adjudicated and seal its records. External funding agencies are obligated to accept university judgments and dismiss cases even when there is ample concerning evidence. Such eye-popping conflicts of interests would be scandalous in business or law. A journalist wrote of an account of *The Baltimore Case*: [10] "You read with a rising sense of despair and outrage, and you finish as if awakening from a nightmare only Kafka could have conceived." This characterization captures essential features of many scientific misconduct investigations, then and now.

In former times, it was assumed that scientists and engineers had assimilated the basic principles of the responsible conduct of research, including the concept of competing interests. Now ethics is taught. Universities require that investigators watch and answer questions about on-line case studies of questionable research practices. The presumption in this training is that only scientists are subject to unethical practices. In contrast, it is presumed that those who administer science do their work in a value-free environment. A young scientist is never told in the face of incessant exhortations to be self-aware and vigilant with respect to misconduct in the laboratory that should she bring forward concerns about questionable science, science that is best unquestioned because it is profitable, she stands a strong chance of being betrayed by her employer, and there is nothing that anyone in government can do to help. Few scientists recognize that the absurd accommodation of competing interests by universities is *required* by federal regulation because these regulations are never revealed in the course of ethics training. Meanwhile, individual investigators must declare their own potential conflicts of interest to federal agencies and publishers many times a year when submitting and reviewing grant proposals and journal articles. Annual conflicts of interest disclosures must be signed and filed by university employees. However, many academic institutions and their representatives are free to carry and exercise massive and systemic conflicts [11]. Such hypocrisy is unsustainable.

An investigator will only learn about our system of scientific misconduct regulation if she is unfortunate enough to discover how truly ineffective they are. Franzen learned these rules the hard way by merely insisting in the course of doing his job that certain facts claimed by his collaborators were mistaken, facts that any of his investigators might easily have checked. But the facts were inconvenient and potentially costly. University leaders, stewards of history, law, and sociology, should not have to be reminded that money makes human beings and the institutions they create less righteous than they might be otherwise.

And, they should not have to be reminded by a chemistry professor like Stefan Franzen. Franzen's expertise is in biophysics, studies at the interface of biology and physics, and as such he was welcomed in a fascinating collaborative project spearheaded by two colleagues who discovered that RNA molecules, principally DNA's helpers in the chemical realization of genetic information, could be "trained," so to speak, to build metal (palladium) particles with catalytic capabilities. This work was a surprise, joining two important yet disparate areas of chemical inquiry. The work generated a high-profile paper in the *Science,* the leading American general science journal. Data therein led to patents that in turn led to major support from entrepreneurs and the administration of another public university. In 2004, professors with a hot result were keenly aware of how to generate capital and interest in the post Bayh-Dole marketplace. Yet, as occurs often in a university, the two inventors were encouraged to team with Franzen to increase their perceived competitiveness in the funding competition announced by a private foundation of an external funding agency. Franzen joined the project after the foundational publications appeared, but for him, it was new territory that he greeted with enthusiasm familiar to most scientists embarking on a quest to understand a new, yet unexplained, phenomenon.

The future was bright until Franzen was notified by a concerned PhD student that the palladium particles did not seem to have much palladium in them. In fact, they were 95% carbon. As a newcomer invited to join the project in order to help gain private funding, Franzen was not privy to the original data. While trying to understand the inconsistent results he noticed another loose thread. He pulled it. For the RNA-mediated metal synthesis to work, it had to work in water, RNA's natural milieu in which it can take on structures encoded in its sequence of building blocks. An aqueous solution was specified in the groundbreaking *Science* publication. But the source of the palladium metal was a compound that did not dissolve in water. If the palladium source did not dissolve in water, the palladium atoms could not be accessed by the RNA, the potentially lucrative process could not work, and the company based on the process would have little value. It seemed that no metal particles had been produced at all. It was the responsibility of his collaborators to check, just as Temin said. Perhaps they did. However, they nevertheless continued to insist that they had invented an amazing transformation that took place in water and aggressively tried to dissuade Franzen from saying otherwise. One obvious issue in dispute was a simple fact – the solubility of a chemical compound – that any high school science student easily could have checked. This difference of opinion about a fact morphed into the longest scientific misconduct investigation in the history of the National Science Foundation (NSF), a sponsor of the questionable research that took 9 years to resolve in Franzen's favor. The length of the process undoubtedly had to do with the fact that according to the Code of Federal Regulations that governs the NSF "Awardee institutions bear primary responsibility for prevention and detection of research misconduct and for the inquiry, investigation, and adjudication of alleged research misconduct." [12] Franzen's one-time collaborators were employed at two public universities, one of which was threatened with legal reprisals by a company in which the other held an equity state. It was in the best financial interests of both institutions to allow Franzen to twist in the wind by disavowing his concerns. As long as universities are able to disingenuously argue that there is legitimate scientific debate surrounding an unambiguous, easily demonstrable error, it can continue to collect money from its financial entanglements or avoid reprisals from a rigorous and transparent examination of the facts. In this case, any administrator or inquiry panel participant could have tried to dissolve the palladium source in water and thus saved many from a painful and protracted struggle. Facts can be constraining if the outcome of an investigation is predetermined.

The Science Bubble is based on Franzen's experience with colleagues who created a fantasy based on easily falsified chemistry, but the book goes far beyond one case – the so-called hexagon case – to examine the general failings of the research misconduct system to properly adjudicate falsification and fabrication of research. The narrative is documented with first-hand evidence as well as that from other cases in the scientific literature. The point of this book is not merely to tell a story about falsification, but to use the evidence to illustrate the ways in which universities have failed to live up to their responsibility, and to their promise that many in society continue rely on.

The characters that are described in the case study are neither heroes nor villains. They are archetypes. There are scientists who were not proactive in the face of legitimate concerns, Baltimore types. And, as the social psychologist Jennifer Crocker makes clear, we all run the risk of not being sufficiently vigorous self-critics. However, once there is a hesitant suggestion that something clearly wrong could be right, then it becomes easier and easier to support subsequent insistences, even if the next step in the defense of the indefensible has become more incredible:

> "Every minor transgression – dropping an inconvenient data point, or failing to give credit where it is due [or in this case refusing to acknowledge a mistake] – creates a threat to self-image. The perpetrators are forced to ask themselves: am I really that sort of person? Then, to avoid the discomfort of this threat, they rationalize and justify their way out, until their behaviour feels comfortable and right. This makes the next transgression seem not only easier, but even morally correct." [13]

The perpetrators in the hexagon case were human and burdened with the psychological limitations that are the birthright of us all.

What kind of character is Stefan Franzen? Franzen recognized that a matter of scientific fact was wrong and said so. In this way, he crossed a line that he did not even know that he was crossing. By suggesting that, *hey, something is not quite right*, by applying the critical outlook that was part of his training as a scientist, by challenging something that looked fishy, a rebuff no different from hundreds of others spread across a career in science, he was, in this instance, unwittingly crossing the line that transformed him from discerning critic into whistleblower. A whistleblower must live with the occupational hazard of counter-accusations.

Franzen was subsequently subject to numerous internal, university-supported inquires and investigations, awash with retaliatory allegations. Despite confidentiality regulations that were supposed to protect whistleblowers, Franzen's case had become gossip in university chemistry departments from coast to coast. None of this makes it easy for a professor to hew to his foremost responsibilities of teaching, research and university service. Nonetheless, Franzen continued to pursue his externally funded scientific interests, disclosed in peer-review publications. His commitment to education led him to create new international programs devoted to science and technology education. This was presumably a necessary distraction from the onslaught of misinformation and hearsay. Opposition to Franzen reached its zenith when he was accused of being a "science bully" for his insistence that demonstrable falsehoods had to be corrected despite the implications for colleagues, journals, companies, and universities administrators. Franzen's unwillingness to concede inspired a short-lived website directed at him called "Stand up 2 Science Bullies," that invited readers to tell their own tales of alleged persecution by colleagues. The website did not create the expected outrage from sympathetic souls and was taken down after more than a year. Nevertheless, a record of it is preserved on the website *Retractionwatch.com* [14]. In addition to the construction of a website targeting a single individual, the informant in an investigation, there were many other behind-the-scenes efforts to discredit Franzen, a few of which are disclosed in this book. The epithet and creative neologism "science bully" falls within the bounds of the so-called "nuts and sluts" characterization of whistleblowers [15].

We can look to history to find others asked to live with knowingly false facts about the world because it was in someone's best interest to accommodate a phony reality. Giordano Bruno was burned at the stake in 1600 for refusing to disavow the wisdom of Copernicus, an opinion that was inconvenient for Roman clerics. Fortunately, Franzen is still with us. Still, he was consumed, if not by fire. He endured a disinformation campaign that was promoted in scientific circles while waiting *for 9 years* until the Inspector General's Office of the NSF finally confirmed that publications were falsified – a justice so torturously slow that anyone should have lost his or her mind. This was not planned. Professor Franzen was merely doing his job as an experimental scientist and in a blink was suddenly in opposition to big organizations. His effort to correct the scientific record in the face of institutional resistance, an effort eventually vindicated, was Herculean. Thousands of hours of correspondence, study, and trips to Washington D.C. were required to produce a judgment that ended up fixing the scientific record but at great personal cost. This is time that Franzen could have used in service to his own science and his doctoral students, but he stepped away from his own interests in order to advance the collective. Franzen's insistence that wrong facts must be removed from the scientific record was a communal act intended to leave the edifice that is science in better shape for all who will come along and carry research forward. That said, Franzen's civic spirit and indefatigability are not sufficient to explain why he undertook the titanic journey described herein. It is possible that had Franzen looked away just once, of the many times he had been asked to do so, he would have accepted that the interests of universities to which he had devoted his career can be at odds with science and truth telling itself. Were Franzen to concede that the university was just like every other corporation, his identity as a university scientist and teacher would suffer a severe attack jeopardizing his every small and daily act in the laboratory and classroom. This was a bargain he could not make. This was a world he did not want to know.

Arguably, the most valuable part of *The Science Bubble* is Franzen's optimistic appraisal of what we can do to ensure that science endures despite the intrusion of financial interests. I am not going to reveal his prescriptions here. These creative and sensible remedies are best discovered as the concluding chapters of the *Bubble* unfold and we would be well advised to consider them in earnest. It may sound like hyperbole to suggest that if we don't fix the system for adjudicating scientific misconduct and restoring public confidence, science will cease. History nevertheless tells of dark ages. There is nothing in the human experience that guarantees that the network of human relationships that is contemporary science must endure. It is the obligation of every scientist to ensure that it does; the program of science is hard, it is multigenerational. Unfortunately, this obligation, in my experience, is not generally recognized or respected by federal regulators or university officers. Ultimately, only scientists can ensure that science endures. Science is the gift that is passed along from teacher to student, carefully, like the penguin's egg in that great movie of endurance [16]. For this reason, we have a right to know when actions of colleagues tear at the fabric of science and threaten our obligation to ensure the survival of a precious practice. When righteous concerns about the veracity of scientific work are investigated, we all have a right to know who said what to whom. We have a right to examine the facts. We have a right to ask questions. Whenever a community is threatened, its members are obliged to clearly understand the threat and its potential consequences. In words made famous by the great Rachael Carson in a different context, wisdom my words have played on here, "The obligation to endure," dear research integrity officers, university administrators, and inspectors general, "gives us the right to know" [17].

New York University
New York City, NY, USA

Bart Kahr

Acknowledgments

This book has benefitted from conversations and critical evaluation by many students colleagues, friends, and family. First, I would like to thank my wife, Maggie, for critically reading the manuscript and providing the crucial advice on how to shorten and stay on point. Second, I would like to thank Chinese and American undergraduate students, Megan Whitney, Zheng Rui, Shufan Cai, YeeShan Leung, and Jing Zhao for their helpful comments and insight. It gave me a fresh perspective. Their comments are reminder to be humble, but also to be aware that writing can convey a different tone than one intends. Third, I would like to thank Alan Sokal, Steven Biggs, and Veljko Dubljevic for helpful discussion and insight on the philosophy of science, Sujit Ghosh for help in presenting elementary Bayesian statistics, and Mariusz Jaskolski for comments on crystallography. Fourth, I would like to thank my father, Fritz Franzen, for critically reading relevant parts of the manuscript to eliminate repetition. His comments and experience helped to navigate that period. Fifth, I owe a debt of gratitude to Dr. Bart Kahr for help in phrasing and focus on the important ethical issues. Bart has become a kindred spirit because he experienced the same legal and regulatory failure of the research misconduct system, but in a dramatic manner. We need to keep in mind that hundreds of serious cases are reported every year. This is not mainstream science, but it is in desperate need of rehabilitation.

Contents

List of Figures

Chapter 1
Evolution in a Test Tube

Science is a search for the truth, that is the effort to understand the world: it involves the rejection of bias, of dogma, of revelation, but not the rejection of morality.

–Linus Pauling

Scientists have high expectations of themselves. Society has come to expect a great deal from scientists because of the role played by technology in the modern world. Science relies on the honor system consistent with Pauling's description of moral behavior in science. Each scientist is expected to tell the whole truth about his or her work as faithfully as possible, whether welcome news of a great discovery or the misery of failed experiments. Despite years of training that reinforces among scientists the use appropriate control experiments and the necessity of examining data critically, the honor system may come undone. There are checks and balances in the system of peer-reviewed research that encourage scientists to lean in the direction of honesty, but there are other forces at work encouraging them to lean the other way. A recent epidemic lack of scientific reproducibility and the rising rate of journal article retractions for malfeasance suggest that the honor system is not up to the task.

The culture of science has evolved dramatically since the end of World War II. Researchers once cosseted in ivory-tower universities have been encouraged to become entrepreneurs and challenged to make science relevant to society. Today's successful scientist must be skilled at marketing research proposals in an increasingly tight science funding climate, savvy about navigating the increasing influence of corporations, and swift to patent inventions. These additional mandates are transforming the honor system into a business ethic in which the risk falls to the buyer, i.e. the funding agency, investor, or journal reader. Since the majority of scientists have dedicated their lives to an evidence-based investigation of a narrowly defined topic, few are equipped to step back and ask whether science as a whole is in danger of succumbing to the financial pressures of the modern university. The pressure to obtain funding combined with the complexity of modern science can lead both mentors and students to push their scientific interpretations and methods into areas where it is possible to either make a mistake or purposefully misrepresent research.

S. Franzen, *University Responsibility for the Adjudication of Research Misconduct*, https://doi.org/10.1007/978-3-030-68063-3_1

What does a misrepresentation look like? Has an appropriate system been put in place to ensure the integrity of scientific publications? These are questions I should have asked myself years ago, but my training, like that of many of my peers, focused on excellence with the expectation that everyone would be scrupulously honest and careful. It was considered obvious that any mistake or self-deception would be quickly spotted by the watchful eyes of the research team. It was taboo to ask whether someone could ever try to fool the group or somehow create a result that looked correct, but actually was a fraud. Because of the assumptions of honesty as the norm, academic journals have been slow to implement enforceable ethical guidelines. Universities typically do even less to implement general procedures that research groups must follow to ensure research ethics.

My training was strong with regard to spotting inconsistencies and logical errors, but it did not in any way prepare me for the possibility that a collaborator would be so sloppy that they could follow a completely erroneous path all the way to publication in a high-profile journal, submit patent applications and obtain legal protection for a fraudulent idea. Like many scientists, I assumed that the scientific community has mechanisms in place to adjudicate and correct scientific fraud. Instead, I learned that the federal regulations place trust universities to adjudicate allegations against their own faculty. The central hypothesis of this book is that federal research misconduct regulations have failed to appropriately consider university conflict of interest in giving them the responsibility to investigate allegations made against their own faculty. I will present facts from research misconduct investigations as evidence in support of this hypothesis. Some of the facts result from personal experience, other examples come from personal communications (with permission) and the third set of examples come from the literature. My hope is that this book will serve as a text for students, a guide for faculty and blueprint for policy makers.

Anyone can make a mistake. Consequently, scientific research has many levels of checks and balances that are supposed to identify mistakes before they make it into print. Published research articles are part of community discussion. Thus, one would imagine that any mistakes that make their way into journal articles would be found out and then retracted. Graduate students and post-doctoral researchers (postdocs) are taught to carefully check their work and examine the work of others with a critical eye to find any small telltale signs that there is a problem in the research. Like many scientists, I believed that the training of the scientific community was sufficient to prevent problems in reproducibility, or worse yet fraud, from becoming widespread in science. I made a hard landing when I discovered that I was wrong. Once I witnessed an ethical problem and began to discern a breakdown in the mechanisms that were supposed to respond to ethical transgressions, I began to investigate the issues. As my own case unfolded with its complicated set of failures to properly adjudicate (what should have been) a simple case of falsification, I came to understand that there is a record retraction rate, thousands of examples of fake peer review, a pervasive lack of reproducibility and sufficient evidence of falsified research results in every field to cause alarm. Beneath the surface of these indicators of a systemic failure in science lies a much greater problem that is masked by confidentiality. The confidentiality clause in federal regulations and university rules is supposed to protect scientists who come forward and to protect the accused in an investigation. The corollary to the main hypothesis of this book is that universities have abused confidentiality regulations to conceal their handling of investigations. Aside from direct experience and observation, there is significant evidence that universities have suppressed investigations, made graduate students the scapegoats, and persecuted whistleblowers. But most of the evidence is hidden from view by the universities which have implemented confidentiality in such a way as to protect themselves rather than the respondents and whistleblowers they are obligated to defend.

Biomedical research and psychology have the highest incidence of research falsification [18]. Stem cell research tops the charts both in sheer numbers, but also in the seriousness of the reported cases. Korea's Hwang Woo Suk, Japan's Haruko Obokata, Sweden's Paolo Macchiarini are among the high-profile cases [19, 20]. The ethical issues in stem cell research range from patient rights or informed consent to honest reporting. Medical research in general has the potential to positively impact human health, which brings prestige and wealth to successful research scientists. Nevertheless, the stakes are very high and unethical practices can harm patients or even result in death. While studies are often enormously complicated, improper statistics and image manipulation can sometimes betray a falsified result. Ironically, wholesale fraud can be more difficult to discover than image manipulation or numerical alterations. A well-executed fraud may leave no telltale clue that the work is completely fabricated. For example, the cases of both biologist Marc Hauser and psychologist, Diederich Stapel, involved dozens of fabricated publications, but neither case set off any alarms for more than a decade [21]. Despite high-profile exposés and books written during the past 35 years, the problem has only grown worse [7–9]. High-profile frauds, lack of reproducibility and retractions are all rampant. The eternal debate about whether there is an increase in ethical lapses misses the point. Fabrication and falsification have been present at some level since the beginning of science. Any level is damaging and arguing about the extent of the damage does us no good. We need to find a way deal effectively with the issue.

My field of research is Physical Chemistry, which involves quantitative measurements and methods that can identify structures and composition of matter with precision. I thought that it would have been impossible for anyone to falsify a research project in such a field of research. It is certainly harder to fabricate a credible result or to falsify data in a way that eludes detection where structure and physical properties are concerned. Under normal circumstances, the supporting data using powerful techniques such as mass spectrometry, nuclear magnetic resonance and X-ray crystallography are sufficiently accurate that there is little doubt regarding the validity of published statements in research on chemical or physical phenomena. Nonetheless, Schön [9] and Taleyarkhan [22] are two notable fraudsters in physical science among many others over the years. There has been an increase in the retraction rate in physical sciences, which is an indicator that falsifications are also on the rise in parallel with other branches of science. One reason may be the trend towards presenting results in a way that will increase their appeal to funding agencies and investors, i.e. the marketing of scientific ideas. The advent of nanotechnology as a branch of science has led to increased use of imaging techniques to present results in chemistry and physics, sometimes with significant omissions in physical characterization of the underlying composition of matter or careful statistical analysis of the yield of a particular type of nanoparticle or structure. This has coincided with decreasing educational standards in many universities as class sizes increase along with use of web-based materials to replace human interaction in the learning process. Web-based education may rival human interaction once it has been fully developed, but the current implementations often fail to give students sufficient insight or challenge them to check their understanding. These changes have resulted in a decline in the level of critical thinking. Papers proliferate repeating similar types of measurements and conclusions regarding potential applications in an echo chamber. Common sense questions are often refuted with glib answers that refer to specialized equipment as though somehow the equipment itself justifies the claim of a new effect or a particular quality of the signal. The complexity of science is used to the detriment of solid research, founded on careful control experiments and critical tests.

The style of writing that controls the narrative to shape a view of reality is possible in fields of research that are sufficiently complex that few readers have the breadth to understand all of the aspects of research and the equipment is sufficiently specialized that data can be presented without full disclosure of how measurements were made. I will systematically consider how two scientists have controlled the narrative to present completely incorrect and falsified results in a manner that has an appearance of validity. I will also complement that description with other case studies known from the scientific literature. In current media coverage of controversial issues, it is common to find biased and cherry-picked reporting, complete with image manipulation. This new type of journalism has become the nightmare of "fake news." Readers are free to choose, which version of the news they trust and in recent years each side has referred to the other as "fake news." The technique of controlling the narrative has crept into scientific discourse and the insidious fake observations and fake data have grown to an alarming extent in the scientific literature.

The hexagon case discussed in this chapter is an example where the result was supposedly proven with specialized measurements. Behind the elaborate technical descriptions of evolutionary chemistry using modified RNA, which I will refer to simply as RNA for the sake of brevity. The idea that RNA could mediate the formation of the hexagonal palladium particles was disproven simply by the fact that the hexagonal particles form spontaneously *without* RNA, as shown by our research in six publications from 2007 to 2013 [23–28]. The authors claimed that the hexagons were composed of "metallic palladium", but the data that were supposed to show this did not actually exist [29]. Instead, the data obtained in two collaborations showed that the hexagons were *more than 95% carbon* and clearly not metallic [30]. The hexagonal particles are nothing more than a kind of carbon "snowflake" formed from a commercially available reagent [26]. Particles with apparent sixfold symmetry rapidly form in solutions of a mixture of water and a solvent called tetrahydrofuran (THF). THF is a clear liquid (solvent) that mixes completely with water at room temperature. To clarify this for the non-chemist, THF-water mixtures are similar to alcohol-water mixtures in the sense that the two liquids are clear, i.e. transparent, and completely miscible. One cannot tell simply by visual inspection that the clear solution is a mixture of different solvents. This detail is important because the two professors from University 1 hid the fact that they used THF since they knew that it would invalidate the entire premise of their idea.

The hexagon case is based on the claim that evolutionary chemistry using modified RNA can create new solid-state materials. However, RNA is not compatible with solvent THF since RNA structure requires a high salt concentration that only water can provide. For the authors, who also considered themselves inventors, the method for producing palladium hexagons was patented as the first example of a general method that they claimed would revolutionize the synthesis of new materials [31]. Behind this elaborate-sounding description was the simple fact that the observed new material was not made by RNA and the material was not a new type of palladium at all, but merely a carbon precipitate [23–28]. In 2004 the two scientists misrepresented the solvent in the prestigious journal *Science*, leading a reader to believe that they had conducted the experiments in water, when in reality half of the solution consisted of another solvent called THF. The misleading description of the conditions served a purpose. If they had admitted that they added the alcohol-like solvent, THF, biologists would have questioned the entire experiment since RNA structure requires water as the solvent. As a scientific collaborator who was invited to join the project after the *Science* paper had already been published, it took me some time to realize that neither the procedures they reported nor the claimed physical measurements were documented properly in the *Science* paper. My research group was not merely invited to join, but I helped the two professors obtain significant funding from the W.M. Keck Foundation and National Science Foundation. As a physical chemist, I felt that my research group had a mandate to understand the experimental evidence behind the claims made in the *Science* paper. Laboratory work is often painstakingly slow. As a newcomer to an existing project in late 2004, it took my research group nearly 11 months to conclude that there was a major error in the presentation in the *Science* paper. After another 10 months of discussion to attempt to correct these problems, in 2006 I was forced to the conclusion that the data were misrepresented in the publication. This conclusion was officially confirmed at the end of a nine-year-long federal investigation. The conclusion of the Office of Inspector General of the NSF was summarized in the 2013 Semi-Annual Report to Congress,

> We investigated an allegation of falsification of research connected with NSF proposals. We concluded, based on a preponderance of the evidence, that the Subjects [from University 1] recklessly falsified research data, and that this act was a significant departure from accepted practices. We recommended that NSF take the following actions: make a finding of research misconduct and send to each of the Subjects a letter of reprimand; require that the Subjects contact the journal in which the falsified data appeared to make a correction; require certifications and assurances for three years; bar the Subjects from serving as a peer reviewer, advisor or consultant for NSF for three years; and require the Subjects to complete a responsible conduct of research training program.

Initially, the scientists may well have fooled themselves. However, when confronted with misrepresentation in their publication, they set out to eliminate any evidence that would expose their self-deception. Fooling oneself is an embarrassing mistake. However, systematically cherry-picking data to justify a self-deception is fraud. Using lawyers to intimidate other scientists who discover such a fabrication should be criminal. Unfortunately, our system tolerates and even protects those who use lawyers to intimidate journal editors, university administrators and other scientists. The use of confidentiality clauses in the research misconduct regulation serves the interests of those who have the means to hire lawyers rather than the interest of an unfettered search for scientific truth. In the hexagon case, the laboratory notebooks provided the crucial evidence of falsification [30, 32–34]. However, I was not permitted to see the written proof in those notebooks until 2 years after the investigation at University 1 was completed. Even though University 1 concluded that the data had been falsified in the *Science* paper, University 1 General Counsel withheld the laboratory notebooks from me without justification. The University had also failed to notify the NSF of the existence of the notebooks. I was not able to independently corroborate my evidence with notebooks until 2 years after the university investigation had concluded, and then only when I hired my own lawyer to press for their release. Despite the simplicity of the scientific case, one needs access to the primary data and records to prove intent as required by federal regulations. By 2010, the University 1 Research Integrity Officer and I were the only people outside the NSF who knew that the NSF OIG was still investigating the case. I was insistent on seeing the evidence myself because I knew that the NSF OIG agents did not know about the evidence in the notebooks. On the other hand, the false statements in the *Science* paper were so obvious that the University 1 investigation committee came to the conclusion that data had been falsified after a 2-year internal investigation *without even having seen the strong evidence in the laboratory notebooks*. Under current research misconduct regulation, a committee conclusion that data were falsified has no impact unless the committee also concludes that falsification was intentional or reckless. Only then is it considered a finding of research misconduct. The federal regulation is so vague that it has been interpreted by everyone to mean that if no research misconduct is found, the falsification or fabrication may be excused, and respondents may be exonerated. When this occurs, all the facts of the case will be kept secret for reasons of confidentiality. The respondents may then claim that they were falsely accused and point the finger at the whistleblower as a vigilante scientist who made a "false allegation". In the current litigious climate, universities will not permit the release of information that would defend the whistleblower for fear of violating confidentiality. In such cases, the faculty who have falsified data can continue to teach and mentor graduate students, receive grant funding and further their careers while the whistleblower may suffer the consequences of having come forward with, what is perceived as, a bad-faith allegation. It appears that the authors of the federal regulation did not consider the all-to-common scenario where scientific peers in an investigation committee are reluctant to find intent, even if they can see that colleagues have falsified data. Yet, the same regulation requires anyone who even suspects falsification or fabrication to come forward and file a statement with the appropriate authority.

Despite the damage to my career I experienced by reporting what I had witnessed, I was one of the lucky whistleblowers. I was a tenured full professor and I was not fired. Moreover, in 2016 the Editor-in-Chief unilaterally retracted the *Science* publication based on the strength of a federal investigation completed in 2015. Although the NSF OIG recommended a finding of research misconduct in 2013, the conclusion reached by the Deputy Director of the NSF modified this finding. The letter of reprimand by the Deputy Director to the subjects stated that,

> consistent with § 18620-3 of the NSF Act, I have determined that from the date that this action becomes final you shall be ineligible for a future award under any NSF supporting program or activity.

Instead of a finding of research misconduct the final ruling was based on the NSF Act of 1950, which sanctioned the respondents for failing to accurately report their research results to the Director of the NSF. Given the litigious proclivity of the respondents and their public threats to appeal any finding of research misconduct under 45 CFR § 689.10(a) of the NSF research misconduct regulation, the explanation for this alternative punishment is given by the statement in the final paragraph of the letter of reprimand.

> Through your counsel, you have requested an opportunity to respond to the allegations made in this matter prior to a final decision being made by NSF. NSF has already given careful consideration to the rebuttal that you submitted in response to the draft OIG report. In addition, because NSF has not made a finding of research misconduct in this case, the appeal rights laid out in 45 CFR§ 689.10(a) are not applicable.

The respondents declared in public that this was an exoneration since it was not a finding of "research misconduct" [35]. The legalistic approach followed by the respondents had served them up to the letter of reprimand in which the NSF used its own Charter to debar them from funding for life. The respondents had been able to avoid being found guilty of research misconduct for 9 years because an academic committee found that they falsified data and violated the norms of the academic community, but lacked the temerity to state that they did so with intent or that they were reckless. Indeed, even though University 1 had established that the respondents falsified data in the investigation report in 2008, the Code of Federal Regulation 45 CFR § 689.1 states that proving falsification of data is only part of the requirement for a finding of research misconduct. One must also establish that the falsification is a "significant departure" from research practices and that it was committed recklessly or with intent. Lacking the finding of intent or recklessness for 9 years the two professors maintained that they had been exonerated by numerous investigations at two universities. This statement had the appearance of truth because no outsider could know that a federal investigation continued for 9 years until the Deputy Director of the NSF finally signed a letter of reprimand debarring the respondents for life.

By 2016 only the respondents and their lawyers could fool themselves into believing that debarment for life imposed by the NSF in the letter of reprimand is less serious than research misconduct, which carries a maximum penalty of debarment for 3 years [36]. In the end, the hexagon case "broke new ground" [37] because of the litigious nature of the respondents, who had stated publicly that they were confident that they could overturn a conviction of research misconduct on appeal [38]. Under research misconduct regulation an appeal will take place in the courts. As with any appeal, the outcome depends only on the technicalities of whether the case was conducted according to the letter of the law. Research misconduct regulations are so fraught with contradictions and vague wording that an appeal would have a good chance of success. In Chap. 9, Judson's account of the appeal in the famous Baltimore case serves as an example of how an appeal can overturn a finding of research misconduct based on strong evidence [7]. In the hexagon case the letter of reprimand by the NSF Directorate removed the option for an appeal by using an unprecedented legal finding, namely that the respondents had violated the NSF Charter [37]. Ironically, one paragraph from the NSF Charter written in 1950 was more effective in adjudicating a contentious case of scientific falsification than the entire NSF research misconduct regulation, which took 20 years to codify in final form (from 1985 to 2005).

The duration of the case and lack of a university correction of the scientific research are both a direct result of extensive legal pressure by the respondents. Despite the obvious falsifications in the hexagon case, the fraud was not publicly recognized for a decade. As of this writing, only one of the five publications containing falsifications has been retracted. More than 12 years after the falsifications were first revealed only the 2004 *Science* paper [29] has been retracted, not by a university but rather as a unilateral action by the Editor-in-Chief of *Science* [36]. In summarizing the aftermath of the federal investigation and retraction of the paper the chemical subject journal *C&E News* reminded readers of the initial assessment of the work in that journal.

> *C&E News* wrote a story about the work [on RNA mediated formation of hexagonal nanoparticles] at the time [in 2004], asking Gerald F. Joyce, of Scripps Research Institute California, to provide expert comment. Joyce was enthusiastic about the results: "We now realize that RNA is adept at synthesizing inorganic materials, and this discovery takes the field of RNA-directed evolution in a whole new direction."
>
> –Stu Borman, Chemical and Engineering News, Feb. 2016, 94, 37–38

Of course, Dr. Joyce would have initially assumed that the 2004 *Science* report was based on accurate statements and that the experiments had been conducted as reported [29]. Accurate reporting is the basis of the entire interconnected web of scientific observations that all scientists depend on. When trust in this framework breaks down, science quickly grinds to a halt since we no longer know what we can rely on. Therefore, scientists do not expect inaccurate or misleading statements to be published in respected journals. It is considered a rare and unfortunate occurrence of scientific fraud when such misrepresentation come to light.

> Asked to comment on whether issues with the 2004 paper have adversely affected related scientific research, [Gerald F.] Joyce [of Scripps Research Institute California] said: "Science is a self-correcting process—even 12 years after the fact."
>
> –Stu Borman, Chemical and Engineering News, Feb. 2016, 94, 37–38

I believe that the antiquated notion of referring to science as "self-correcting" is harmful to the conduct of science. Who precisely is the "self" in the term "self-correcting"? Did the scientists who obtained millions of dollars based on false statements realize their mistake and correct it? On the contrary, their lawyers threatened to sue anyone who published corrections using false allegations of intellectual property violations. Nor did the university that investigated the matter, and concluded that the data were falsified, request a correction of the literature.

There has been no public correction of any kind that explains the falsification in the hexagon case to the scientific community. The retraction by the editor-in-chief of *Science* did not explain the fraud. The articles that we published [23–28, 39, 40], to correct the record were ignored by the *C&E News* article [36] and every other report of the case in scientific magazines [37, 41]. Ironically, reporter Joe Neff did more in an article in a regional newspaper to explain the actual scientific fraud than any science editor [30]. Regarding the question concerning negative consequences for the scientific community, the answer is that there have been serious repercussions for numerous graduate students who obtained PhDs based on falsified data and wasted time and expense by research groups who attempted (and failed) to reproduce the results. There has been a waste of millions of dollars of research funding. There has been no public recognition of the whistleblower's good faith report, as recommended by federal regulations. The request for action by University 1 was clearly made in a letter sent by the Chairman and Ranking member of the House Subcommittee on Investigations and Oversight to the Chancellor of University 1 on July 15, 2014, nearly 8 years after the fraud was first reported to university authorities [33].

Dear [Chancellor of University 1],

The House Committee on Space, Science and Technology has been aware of a long-running case of research fraud, which was supported by a grant from the National Science Foundation (NSF) awarded to your university. Results from the NSF-supported work were published in the May 7, 2004 issue of *Science* magazine and highlighted in a press release from University 1. The fact that the three authors, [...] have moved on from University 1 does not lessen our concern.

We write today to express our dismay that the integrity of the research record for this NSF-sponsored work remains damaged. Our dismay is only heightened by the recent revelations that this practice appears to be wide spread (sic). As of this letter the Library of Congress informs us that there have been 115 first generation citations of the 2004 article in various scientific publications. This number does not include the citations generated by the controversy in the scientific literature on the validity of the original findings. [...]

While the NSF OIG investigation [...] remains open pending a final reply by NSF, given the damage to the scientific record that has already occurred and continues to occur, we would appreciate your responses to the following questions:

1. When University 1 contacted Science magazine in 2009, did it provide them a copy of the University 1 investigation and determination that data was unjustifiable. If not, why not?
2. In light of the recent faculty vote at University 1 and completion of an NSF OIG report with recommendation to penalize [the authors of the Science paper] will you write to Science magazine again to alert them to these series of events, the University's findings, and the NSF OIG's recommendations that the article be retracted and that NSF require three years of "assurances and certifications for each author and bar each author as an NSF reviewer, advisor or consultant" for the same number of years. If not, why not?

3. Title 45 689.4 of the Code of Federal Regulations states that awardee institutions "bear the primary responsibility for prevention and detection of research misconduct", which require them to take "action necessary to ensure the integrity of research [...]". Please explain how University 1 has met its research misconduct responsibilities under federal regulation in the case involving [the authors of the *Science* paper].?

As set forth in the letter from Congress, University 1 had determined that data were falsified in an investigation that concluded in 2008. At the time of the letter, the investigation had continued for another 5 years under the auspices of the NSF OIG who had recommended a finding of research misconduct with the full sanctions available under federal regulation in the Semi-Annual Report to Congress in 2013 [42]. However, the response by the Chancellor of University 1 dismissed the Congressional request for documents and declined to notify the editors, citing the fact that the Deputy Director of the NSF had not made a final determination in the case. Eighteen months later when the Deputy Director made a final determination and the editor of *Science* had unilaterally retracted one of falsified papers, the Chancellor of University 1 still declined to take action to notify the other journal editors. Instead, the university decided to accept 3-year old allegations that my research group did not have the right to study certain samples cited in one of our articles written to correct the research record [25]. It was common knowledge in the Office of Sponsored Programs at University 1 that we had obtained joint grant funding with the respondents. That was how I came to be involved in the matter to begin with. Nonetheless, the research misconduct inquiry targeting my collaborative research lasted 18 months and caused significant stress precisely at a time when University 1 should have announced the end of the hexagon case. The research misconduct inquiry ended in a private admission by the university that the allegations were tainted by "retaliatory animus".

The stakes are very high for a person who identifies a case of scientific fraud that involves professors or senior scientists. The whistleblower, as the informant is called in federal regulations and university rules, is often also considered a suspect in such cases. Fellow scientists are likely to be skeptical of an allegation that a colleague of theirs has falsified or fabricated data. Scientific whistleblowers are subject to many of the same prejudices as whistleblowers in other walks of life. They have been ostracized, lost their employment, and been investigated as result of coming forward to report fraud or other wrongdoing. We can begin to understand this problem by considering an actual case where the scientific community, university administrators, journal editors and collaborators were misled.

1.1 Loopholes in Peer Review and Federal Regulations Enable Publication of Falsified Data

In 2004, an article entitled "RNA-Mediated Metal-Metal Bond Formation in the Synthesis of Hexagonal Palladium Nanoparticles" was published in *Science* magazine. The article reported a remarkable ability of the biological polymer, RNA, to control the assembly of palladium metal atoms into micron-sized hexagonal-shaped metallic particles [29]. The laboratory notebooks of the graduate student show that the research was done at University 1 up to 2007 when the respondents moved to University 2 [43]. From that time forward the respondents research in the field was directed at defending the *Science* paper. Further publications include the *Journal of the American Chemical Society* (*JACS*) in 2005 [44] and *Langmuir* in 2006 [45]. The two scientists had combined hot research topics RNA catalysis, materials synthesis and nanotechnology to grab the attention of university leaders, funding program officers and investors. I became involved in the project at University 1 when the two professors and my Dean of Research asked me to join a collaboration to compete for a $1 million W.M. Keck grant in the summer of 2004. This grant was funded on the first of January 2005. Once my research group joined the effort, I applied for and received collaborative funding with the two scientists from the National Science Foundation to study the RNA structure. By mid-2005 they had received grants totaling more than $2.3 million, most of which I did not know about and was never informed. Although I joined the collaboration with an open mind and with the hope of making a significant contribution, within a few months of starting work on the research my group began to find inconsistencies in the reported data. Although I did not publish jointly with either of the authors on any of this work, I did have the responsibility to report to the funding agencies as the recipient of two collaborative grants. I also had the responsibility to mentor graduate students and a post-doc involved in the work. Once we recognized that the two professors did not plan to correct their own work my research group published numerous scientific journal articles to expose the fallacies in the hexagon research [24–28, 39, 40, 46]. Looking back on a series of unfortunate events begs the question: how did this research get published in one of the most prestigious journals in the world of science?

The submitted *Science* article tapped into the interest in RNA catalysis, materials synthesis and nanoparticles in the fields of biochemistry, materials science and chemistry, respectively, with language designed to grab the attention of the scientific community. The authors used the complexity of their concept of "evolutionary chemistry" as a loophole to circumvent criticism. The first loophole used to escape criticism was that the work was patented and counted as intellectual property. This meant that they could deny others access to primary data, samples or any other information using the excuse that it was proprietary information. Later when we identified problems with the research, they used this reasoning even against my group, their collaborator and co-funded research group. A second loophole that permitted the two scientists to evade criticism was that the research comprised such a wide range of expertise that few individual reviewers would feel qualified to discount it. A materials scientist typically does not know much about RNA and biological techniques and vice versa. It is of general interest to understand how the two scientists used complexity and finely tuned phrasing to mislead readers and presumably reviewers as well. They implied that RNA was catalyst, but did not use the word catalyst, which gave them plausible deniability if any scientist began to ask what exactly they meant by catalyst. Science is supposed to be clearly written so others can understand a result, reproduce it and build on it. By contrast, the technique used by the authors from the beginning was to present the idea as so complex that only *they* could carry out the experiments successfully. Again, this was implied in their statements and not stated in public since saying this directly would betray how far they had strayed from the scientific method. But when they were investigated they used this argument repeatedly to claim that the committee did not have the expertise to judge the science. If the investigation committee could not understand the science how could they justifiably find data had been falsified? It is an arrogant and absurd argument, but it did create sufficient doubt that the committee did not feel that they could make a judgement about the intent of the scientists, despite the obvious nature of the falsifications.

To understand the evidence supporting my claims, we have to delve into the concepts and language that they used. I know that I risk alienating the non-scientist by describing the details of the research, but my aim is to explain the ideas in plain language so that any educated reader can understand the misrepresentation of the facts and falsification of data. A catalyst accelerates a chemical reaction and remains unchanged in the process. The authors implied that RNA was a catalyst, but they never defined a mechanism. Since RNA had already gained acceptance as a catalyst in biology the statement that it could catalyze "inorganic-particle formation" may seem to be more justified, in spite of the fact that they presented no evidence to support the claim. It is crucial to differentiate the catalysis of palladium atoms to form a particle (such as reported in *Science*) from the catalysis of RNA to cut and splice itself. When RNA acts as a catalyst in biology, it uses part of its own structure to cleave its own backbone that binds individual units of RNA into a long polymer. This occurs in solutions of water with salts. By contrast, the palladium-containing reagent used in the *Science* paper to deliver palladium atoms to RNA does not dissolve in water. To be precise the palladium reagent used by the two scientists does not dissolve in water to any measurable extent as stated in the Materials Safety Data Sheet (MSDS), the official safety document required to be available in any laboratory where the compound is used. Yet, the *Science* paper states that the experiments were carried out in "aqueous" solution, which means a solution that consists of water. In using the word "aqueous" it is the obligation of the scientist to state explicitly if any liquid or solid has been added to the water. In the absence of any additive "aqueous" means pure water. The two scientists wrote that they discovered a process to make palladium nanoparticles in "aqueous" solution using a substance that does not dissolve in water. How could such an obvious discrepancy have been missed by the reviewers?

The research described in the *Science* paper is at the intersection of biochemistry, organic chemistry, and materials science. Mistakes in the review process are more likely in multi-disciplinary research projects since reviewers may be reluctant to judge areas outside their expertise. Scientists who read the article and suspect that there may be a problem will be less likely to make a definite conclusion for fear that they lack the expertise to completely understand the issues. Artificially inflated complexity creates a loophole in the review process that can permit ethically compromised publications to slip through in a manner that is analogous to tax loopholes, which permit certain wealthy individuals to avoid paying taxes. People who successfully use loopholes may be tempted go one step further and resort to illegal means. Peer review can fail for other reasons such as backroom deals with editors that permit reviewers to engage in mutual arrangements. These are a clear violation of ethical norms in science.

I did not expect any of these issues when I accepted the challenge to work on the hexagon project. I believed the two professors when they described a process for systematically selecting a small number of RNA sequences from solution of many trillions of different RNA sequences. Much of what they reported looks like other studies that use "RNA selections". In this type of research, RNA is synthesized in the laboratory using a relatively straightforward procedure. The two scientists followed published procedures for creating RNA catalysts and then claimed that they could use these methods to synthesize new metallic or solid materials. They never attempted to explain the mechanism whereby the RNA would be able to mediate the formation of metallic particles with a hexagonal shape on the length scale of approximately one micron, at least one hundred times larger than the RNA itself. The selection cycle is shown in Fig. 1.1, which is reproduced (with permission) from the journal *Science*. Using synthetic methods in an automated DNA synthesizer one can create approximately one hundred trillion different DNA sequences in the same test tube. The huge number of DNA sequences can be copied into RNA sequences in a test tube using a natural enzyme. This large number of sequences is called an RNA library. Just as we may select a few good books from a library, the chemical "selection" is designed to find a few desirable RNA sequences from the library of sequences. Of course, this method presupposes that one has a good method to select appropriate sequences capable of carrying out a desired process.

The difficulty in designing a good selection turns out to be one of the weaknesses of the idea as executed in the *Science* paper. In nature, genetic diversity leads to selection by the Darwinian idea of survival of the fittest. In living organisms, the selection for advantageous adaptations is the ability to reproduce. By contrast, most RNA selections used in chemistry laboratories are based on the binding of the RNA to a protein "target" [47, 48]. The binding selection is based on the idea that one can separate bound RNA to the target from unbound RNA, which is still in solution. Figure 1.1 shows this concept. Green-colored RNA binds to a target and stays bound after rinsing away the red-colored unbound sequences. The cycle is completed by making a DNA copy of the selected RNA, amplifying the DNA (making millions of copies of it) and then copying that sequence back to RNA. One could call this approach "evolution in a test tube". Prior to the claims of a new kind of "evolutionary chemistry" many other scientists had demonstrated the value of RNA selections for binding affinity to create new pharmaceuticals or new biological markers for cell biology studies [48–60].

RNA has a range of specific functions in biology, but before the report in *Science*, RNA had never been reported to make metal particles. If the concept of the *Science* paper had been described accurately and had worked as indicated, it would have been a breakthrough in how we think about the discovery of new materials. However, none of the commentary on this concept addressed the question; how could this

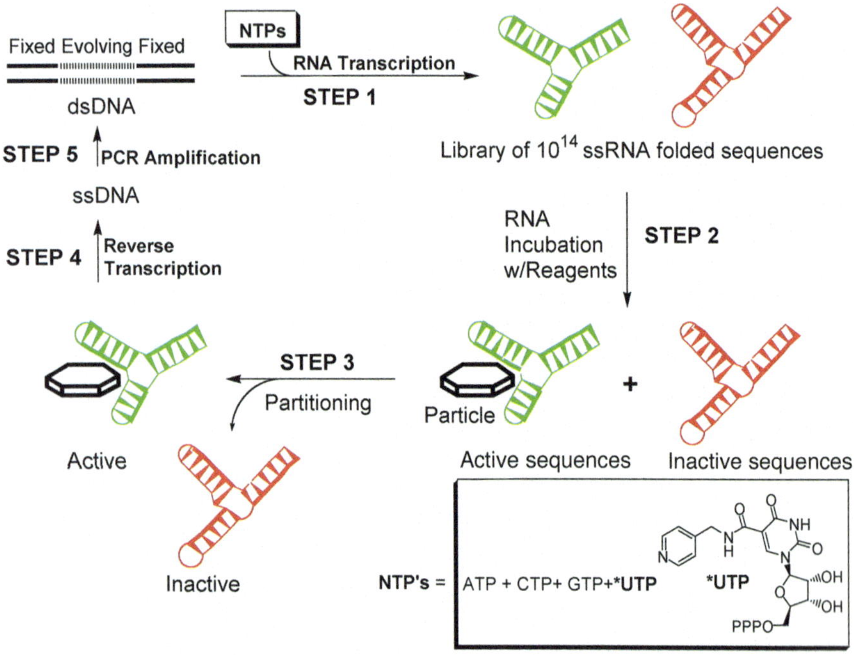

Scheme 1. Steps of the RNA in vitro selection cycle.

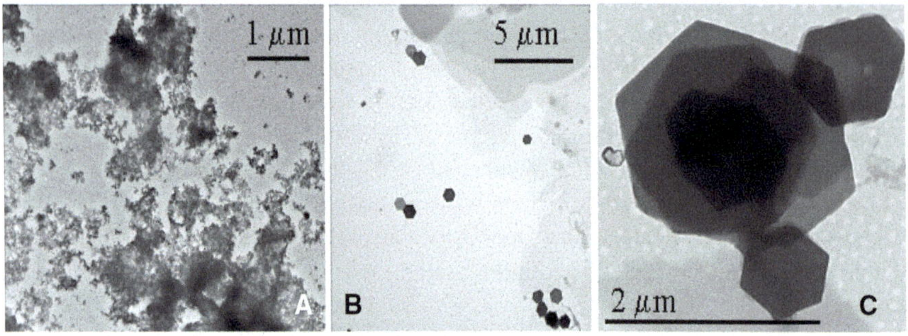

Fig. 1.1 Reproduction of Fig. 1.1 and Scheme 1 from the, now retracted, 2004 *Science* paper

possibly work as described? Unless we understand what has been observed, there is a danger that the result is simply an artifact or worse that it was never accurately described in the first place. An optimist might say that some of the RNA sequences with the desired ability will surely be discovered because of the power of evolution. With hundreds of trillions of RNA sequences, one might imagine that surely one of those sequences would have some special property and would be selected. This optimistic view, which was put into grant proposals by the two scientists, ignores the fact that the chemical nature of RNA is quite limited. Despite the power of "evolutionary chemistry" it is entirely possible that none of the sequences can make a new material. Unbridled optimism can easily become wishful thinking. One must be careful to conduct all experiments under appropriate conditions for chemical activity and solubility. A realist will observe that these conditions are quite limited in the case of RNA [61]. For example, RNA can be dissolved in water, but if one were to mix other chemical solvents, such as alcohol in a high proportion relative to water, the RNA would denature. It would lose its structure and aggregate. There are other potential problems with the idea of RNA as an agent for materials synthesis. Even in water, RNA is fragile in the sense that many common metal ions (i.e. iron, cobalt, lead etc.) can catalyze RNA "cutting", which means that they can bind to RNA to facilitate processes that breaks the RNA sequence into fragments. In many respects, RNA is an odd choice for a new molecule to make the next generation of new materials since RNA is itself an extremely fragile molecule easily degraded by trace amounts of metals or proteins.

In biology, RNA is a catalyst that works mainly to cut and paste other RNA molecules involved in translating the information from DNA into proteins that can carry out the activities of a living cell [62, 63]. For many years the prevailing view in biology was that a given DNA sequence codes uniquely for a protein. Prior to 1985, it was understood that RNA primarily functions as a messenger that transports a copy of the information in DNA to the ribosome, the factory that makes proteins in the cell. The contemporary view is that RNA controls the cutting and splicing of messenger RNA to produce multiple possible sequences from one DNA sequence. This insight explains how one DNA sequence can code for multiple proteins [64]. This is important because the sequencing of the human genome revealed that each human possesses less than 30,000 genes, which is apparently fewer than the number of proteins needed to explain human growth and development. Since RNA can act as a catalyst that can cut and paste other RNAs, there is a apparently a complex communication and feedback network within the cell that leads to different RNA sequences depending on the stage of human development [61]. RNA became a hot topic because the discovery that RNA can act as a catalyst has turned out to be crucial to understanding human biology. Starting in the 1990s the apparently wide range of biological abilities led chemists to use RNA and modified RNA for new applications that were previously unimaginable, such as catalyzing non-biological chemical reactions. The idea of going one step further and creating new materials is creative, but it was not grounded in careful science. The scientific issues that reveal the problems are among the easier types of questions to understand. Are the hexagonal particles made of palladium? Did the scientists provide any data showing the chemical composition of the hexagons? Does the starting material used by the two scientists dissolve in water? Did they report any other solvent or addition to water? The answer to each of these questions is "no". This was not a hard case for a scientist to understand once facts were assembled and the misrepresentations in their publication were exposed. The more difficult question to answer is: why did it take 9 years for universities and federal agencies to issue a final conclusion that the data were falsified?

These is no simple answer to the question "why did this case take so long to adjudicate?" One reason is that the two scientists moved from University 1 to University 2 in 2007, which was after the inquiry phase had already been completed and a full investigation was recommended. Although the authors of the article, two professors and their graduate student, conducted the research published in the journals at University 1 from 2002–2006, the two professors used their new address at University 2 to claim that research had actually been done at University 2 rather than University 1. They used this distance to claim that I was not a collaborator and did not have any right to the samples or data that I had published. The research reported in the journal *Scanning* in collaboration with the group of Dr. Jim De Yoreo of Lawrence Berkeley National Laboratory was published in 2008 without any mention of University 1 despite the fact that graduate student's laboratory notebook showed clearly that the samples were made at University 1 [65]. Normally, one does falsify the address where research is conducted. But in this case the two scientists did exactly that to avoid the being in the jurisdiction of University 1 where their research was being investigated. Indeed, the investigation committee at University 1 did not believe that the research reported in the *Scanning* paper was relevant because the work was not done at NC State. They never checked the laboratory notebooks to find out that the samples studied in the *Scanning* paper were indeed processed by the graduate student at NC State.

There are some famous cases where jurisdiction has played a crucial role in delaying investigations, such as the pig farmer near Vancouver, Canada who managed to commit fourteen murders in 4 years during which the Vancouver Police tried to convince the Royal Mounted Police to take the case seriously [66]. The Vancouver Police did not have jurisdiction even though the murdered women were all from Vancouver and Mounties apparently did not see the urgency in the case since they were not from Vancouver and many of the victims were poor or indigent women. As in the sordid case of the Canadian serial murderer, the fact that there were two jurisdictions became one more loophole that the two scientists used to avoid the facts from becoming known to the scientific community. It defies reason that universities would refuse to cooperate or to consider the validity of clear evidence of scientific falsification. I personally witnessed how the flawed definition of jurisdiction has contributed to a crisis in the adjudication of falsification of research. Federal regulations designate universities as the parties responsible for investigating allegations made against their own faculty. The hexagon case is a starting point, but once I began to research the topic, I came to understand how the resulting institutional conflict of interest has weakened the integrity of science. Of course, I am aware that these are strong statements, which must be supported by evidence. There is evidence from many sources. The hexagon case provides the most detailed look to date at the role played by jurisdiction, confidentiality, intellectual property, peer review and institutional conflict of interest in the adjudication of falsification and fabrication.

1.2 A Flawed Experimental Design: Mutually Incompatible Solutions

Over the years, five scientists have told me that the Science paper is so obviously wrong that they could tell just by reading it. Unfortunately, none of those scientists came forward to say this to me until long after the contradictions in the paper had been published by our research group [24]. Only my collaborator, Professor Ryszard Adamiak of the Polish Academy of Sciences, spoke up and expressed his grave concerns about the paper in the summer of 2005. Dr. Adamiak is a world expert in RNA structure [60, 67–77]. Since we had proposed an NSF-funded project in collaboration with his research group I met with him to discuss the grant proposal. In a private conversation he told me that he was reluctant to be part of any project to study the RNA structure in the hexagon project since, as he put it, "the conditions described in the *Science* paper are impossible." He had difficulty explaining the reasons that he had for thinking that it was impossible. He made this warning 5 months before we obtained the first experimental results that contradicted the published account. Dr. Adamiak was able to express the problems with the *Science* paper even though it omits the crucial information that the experiments were conducted in a solvent mixture and not in pure water. Dr. Adamiak was expert in both RNA structure and synthetic chemistry so he could see that RNA required water (actually a buffered salt solution) as a solvent and the palladium-containing compound required a non-aqueous solvent such as THF. He surmised that the authors of the *Science* paper must have omitted something important because the conditions as written are impossible. How could one combine these two mutually exclusive conditions without serious risk of an artifact?

The abstract of the 2004 *Science* article reads as follows:

> RNA sequences have been discovered that mediate the growth of hexagonal palladium nanoparticles. In vitro selection techniques were used to evolve an initial library of ~10^{14} unique RNA sequences through eight cycles of selection to yield several active sequence families. Of the five families, all representative members could form crystalline hexagonal palladium platelets. The palladium particle growth occurred in aqueous solution at ambient temperature, without any endogenous reducing agent, and at low concentrations of metal precursor (100 micromolar). Relative to metal precursor, the RNA concentration was significantly lower (1 micromolar), yet micrometer-size crystalline hexagonal palladium particles were formed rapidly (7.5 to 1 minutes).

The claims made in the abstract are remarkable. For a molecule such as RNA to create a well-defined hexagonal shape, it must somehow pull atoms from a palladium-containing precursor and arrange them in a hexagonal shape that is about one hundred times larger than the RNA itself. A single atom of palladium does not behave like a metal, but a cluster of palladium atoms becomes metallic at some critical size, which is typically on the nanometer length scale. The process described in the abstract sounds so improbable that one would imagine that anyone reviewing the paper would take a long look at the control experiments, the proof that the particle was palladium and many other details. I was not a reviewer, of course, so I assumed that evidence had been presented and carefully evaluated prior to publication. I admit that I was not as critical of the paper as I would have been if I had been asked to review it. But I would say that this is normal since scientists are supposed to have some trust in the review process, not to mention their colleagues. For example, the *Science* paper stated that electron diffraction had provided proof that the particles were "crystalline metallic palladium" [29]. I trusted the authors and did not ask to see the data when I became involved in the project in late 2004. I did not discover until nearly 18 months later that neither the electron diffraction measurement nor the solvent conditions were accurately reported in the *Science* paper.

An obvious control experiment to test for the role of RNA in making the hexagonal particles would be to test the same solutions without using RNA (RNA-free control experiment). Although they did not conduct the RNA-free control, the authors claimed that they conducted a control experiment using a polymer instead of RNA. But as it turns out the polymer was not soluble in water, so they conducted that experiment in chloroform. The properties of chloroform are different from water. The authors failed to mention the use of chloroform in the *Science* paper and I learned that they had used chloroform only when I finally got access to the laboratory notebooks of the graduate student, 2 years after the university investigation had concluded. Obviously, such an experiment is not a proper control since it is not compatible with RNA in water. It is false comparison. Scientists often call such an incongruous juxtaposition an "apples to oranges" comparison. Normally, the reviewers are supposed to ask to see the data or evidence that the control had worked and in so doing they would notice such details as the use of the correct solvent or whether the data even existed. However, when the paper is published and the scientists have a good reputation, many scientists will give the publication the benefit of the doubt. The average reader of *Science* would only have seen a remarkable result apparently validated by peer review. The fact that palladium atoms can combine to form palladium metal is not remarkable, but the reported shape control and the rate of formation of the metallic particle are crucial to the notion that RNA can create particles with control of their shape.

Figure 1.2 shows the lack of solubility of the palladium-containing reagent in water. This is the reason that the two scientists were forced to use an "organic" (mostly carbon) solvent that mixes completely with water. The solvent they chose is THF, which is a clear liquid that mixes completely with water just like common alcohol. Does it matter that the *Science* paper reported an "aqueous solution"? Is it just a detail or a small mistake that the authors actually used 50% THF and 50% water and failed to state this in the *Science* paper? Let us examine an example where the amount of alcohol is misrepresented to determine whether this misleading statement has any consequence. In proper chemical nomenclature, one could call a mixture of water and ethanol "aqueous" only if one specifies the percentage of ethanol in the solution. If someone offers gin that is "80 proof" the solution is 40% alcohol by volume. If the purveyor of the gin calls it an "aqueous solution" without specifying the presence of alcohol, then they are misrepresenting an alcoholic drink as water. If someone were to drink the "water" without realizing it contained alcohol (perhaps because it contained fruit juice to mask the alcohol) then that person could

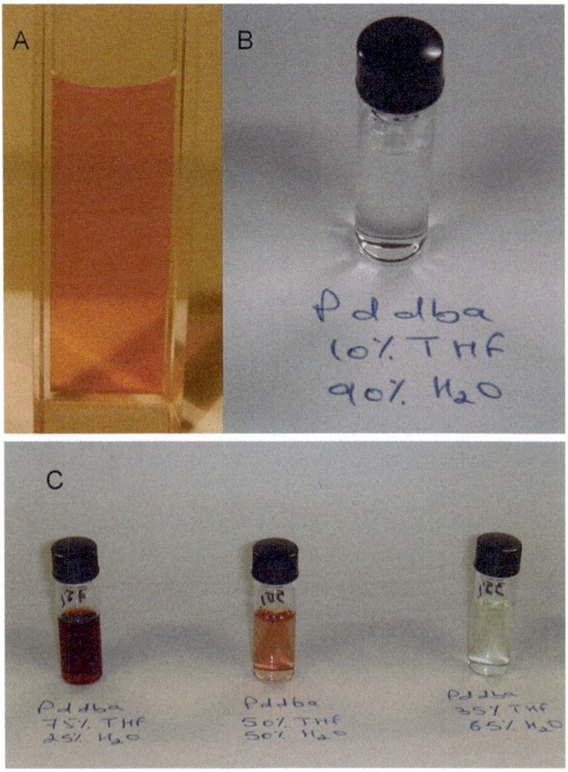

Fig. 1.2 Solubility of the palladium-containing reagent. The figure **a** shows the only reported image by the two former University 1 professors compared to the results obtained by the manufacturer. (**a**) Copy of the picture taken for Fig. 1.1 of the *JACS* Correction [80] and the *Science* e-letter Response by the original authors. This is purported to be a solution of 400 micromolar in the palladium-containing reagent in 10% THF/90% H_2O. (**b**) A solution of 400 micromolar palladium-containing reagent in 10% THF/90% H_2O after filtering the precipitate prepared by the manufacturer STREM Chemical Co. (**c**) Representative solutions of the same palladium-containing reagent in 75%, 50% and 35% THF (right to left). The solutions shown in pictures (**b**) and (**c**) are part of a solubility study conducted by STREM Chemical Co., a leading manufacturer of the palladium-containing reagent

well become intoxicated without his or her knowledge. This is clearly unethical and it would be illegal to do this in a place of business. In same vein, to claim that one conducts an experiment in an "aqueous solution" and to fail to mention the presence of solvent such as THF is a falsification. In this case, the reason for the fraud was to hide the fact that THF was present at high percentage since no reviewer would have approved of a study of RNA in 50% THF and 50% water. RNA structure would be strongly modified in such a solution and it would raise questions about whether the palladium-containing compound was precipitating. Indeed, it was precipitating or falling out of solution like snowflakes, but snowflakes made of a carbon material sinking to the bottom of the container.

This somewhat tedious discussion of solubility is useful because it reveals how a scientific author can distort the facts and prevent the misrepresentation from being exposed. The style of writing used by the authors resembles biased journalism designed to control the narrative in partisan news reporting. Selective presentation can be used give the appearance of a basis in fact. For example, in a subsequent article published in *JACS* in 2005 the two professors reported preparation of hexagons using the same process reported in *Science*. However, the solvent they reported was a mixture of a solvent called tetrahydrofuran (THF) and water, 5% THF and 95% water by volume [44]. This statement made in the small font at the end of the *JACS* article does not agree with the *Science* paper, which makes no mention of THF. The two professors would later claim that this statement was "correction" of their original omission of any mention of THF in *Science*. It is clearly a contradiction, but in what sense is it a correction? The *JACS* paper does not mention the previous omission. How would any reader connect the reporting of solvent conditions with the previous publication unless there were some discussion of the issue. Of course, in reality it was not a correction. The federal investigation revealed that the laboratory notebooks reported use of 50% water/ 50% THF in every experiment, except the first one in 2002, which was a failure. The two professors took great pains to prevent anyone from seeing the notebooks or knowing that they had used these conditions. Yet, any scientist could have checked their conditions very easily as we did when we began to suspect that something was wrong. After spending 1 h in their lab I was had enough evidence to prove that the claims of both *Science* and *JACS* papers are impossible.

Statements concerning the solubility of the palladium-containing reagent are easy to verify since its solution has a purple color in THF solutions [78]. Figure 1.2 shows how the color of varies for mixtures of the solvents THF and water containing various concentrations of the palladium-containing compound. Figure 1.2c shows a solution with a deep purple color in a 75% THF/25% water solution. A 50% THF/50% water solution only has a faint color, which tells us that the solubility is already quite low. In this solution, we would still say that the palladium-containing reagent is "soluble". The vial furthest to the right in Fig. 1.2c shows only a hint of color, which tells as that 35% THF/65% water hardly dissolves the compound at all. A 10% THF/90% water solution shown in Fig. 1.2b is essentially clear. In common parlance, a chemist would say that the palladium-containing reagent is "insoluble" in 10% THF. Even a sensitive instrument called a spectrophotometer cannot measure any absorption due to the palladium-containing reagent under these conditions as reported in the *Journal of Chemical Education* [40]. All of this seems so simple and so obvious that one might wonder how it could become a scientific dispute. However, the two former University 1 professors have shown the same solution in the cuvette in Fig. 1.2a in three separate publications and each time they have claimed that it is a solution of the palladium-containing reagent in 10% THF/90% water [79–81]. This is an unusual claim to make since there is THF "inflation" in their work. They started by saying that there was no THF (*Science*) [29], then 5% THF (*JACS*) [44], then in their "corrections" [79–81] they said that they had used 10% THF. These claims are inconsistent with data obtained from the manufacturer of the palladium-containing compound shown in Fig. 1.2b. The data from STREM Chemical Co. confirm that a 10% THF/90% water solution is clear, which means that it does contain a detectable amount of the palladium-containing compound. Instead, the color of the solution Fig. 1.2a somewhat resembles the solution of 50% THF/50% water in Fig. 1.2c, which is consistent with 50% THF/50% water reported in the laboratory notebook [30].

The detailed discussion of the solution conditions may sound picayune, but a telltale clue that something is not right often leads to the discovery of further discrepancies. In fact, experienced reviewers often look for telltale inconsistencies to help them in quickly formulating an opinion about a paper that they are asked to review. Clearly, the reviewers of the *Science* paper should have asked for more information and been alerted by the inconsistency.

1.3 Electron Microscopy Data Show that the Hexagons Are *Not* "Metallic Palladium"

The two professors from University 1 published three articles in separate journals, *Science* (2004) [29], *JACS* (2005) [44] and *Langmuir* (2006) [45] without ever producing any electron diffraction data to support the claim that the particles were composed of metallic palladium. Our first significant indication that something was very wrong with the claims of the Science paper came at the end of 2005 when we concluded a series of electron microscopy experiments on the structure and composition of the hexagons. We found that they were composed of carbon and had no discernable structure, certainly not a metallic structure. Despite the obvious inconsistencies and refusal by the two University 1 professors to provide data in accord with the publishing agreement, the editors of *Science* denied us the opportunity to publish a Technical Comment, which is a short commentary that points out potentially serious problems in published work. Therefore, we expanded our Technical Comment and published an article in *JACS* in 2007 that described a series of experiments on the hexagons showing that they were *not* composed of palladium [24].

In 2008 we were permitted to publish a short e-letter on the *Science* website. Although the editors of *Science* rigorously restricted us to a maximum of 200 words and no figures [27], they permitted the two professors to simultaneously publish a response immediately following ours with 500 words and two figures [79]. In the e-letter the two professors from University 1 claimed that they had corrected their earlier omission by publishing electron diffraction from the hexagonal particles [29]. The data shown in the e-letter was obtained by the graduate student as a practice trial prior to taking a course in electron microscopy. In her laboratory notebook, she wrote that she could not obtain an image of the sample because it had aggregated. She stated that these data were not usable and she did not report any calibration since she had not yet studied the method in a class. Neither of the two professors had ever studied electron microscopy either. Thus, the purported "correction" in the e-letter shows electron diffraction data, but without an image, i.e. picture, of the sample, without indexing, and with a knowledge that the sample had aggregated. This is not accepted practice in the research community.

The presentation of electron diffraction without any other information to permit interpretation of what the sample actually was, or without quantitative analysis of the structure of the metal should not have been acceptable for publication even in an e-letter. Having such data is a bit like having the X-ray of a patient's teeth without knowing who the patient is. A picture of the patient's face in the file with the X-ray image of the teeth would permit the proper association to be made. The editors of Science and particularly the reviewers should have required that the authors provide an image of the sample they studied. That is standard practice. This omission had an enormous impact on the course of decisions made by university administrations and the judgement of many scientists who are not experts in electron microscopy. Since the editors of *Science* published the diffraction data, other scientists and university administrators who had no expertise at all were under the impression that the data shown in the e-letter actually was evidence that the particles were made of palladium. The editor's acceptance of this flawed data prolonged the apparent "debate" and all of the behind-the-scenes actions to thwart publication of corrections to this misrepresented research.

To anyone in the scientific community who was interested in these questions, this series of papers would have looked like a debate in the scientific literature. To be sure, there were some unusual features of this apparent dispute. It is not often in a debate that two research groups publish opposing viewpoints based on data obtained on the *same* samples. It not likely that many scientists took the time to understand this nuance. There appeared to be no resolution to the debate since in the absence of confirmation by another group such debates tend to end in stalemate, even when one side has overwhelming evidence. The reason for a stalemate is that observers will be concerned that the matter is a result of different conditions in different laboratories. Indeed, it often happens that scientists disagree, but then find different conditions in their respective laboratories to explain apparently contradictory observations. However, the fact that conditions can differ has been used as a smokescreen to deflect scientific criticism as well.

1.4 Scientific Corroboration: The Gold Standard

In 2013 the Franzen and De Yoreo research groups jointly published paper in the journal *Particle & Particle Systems Characterization (Particle)* with cover art shown in Fig. 1.3 [23]. The paper in the journal *Particle* was corroboration with Dr. De Yoreo's group at Lawrence Berkeley National Laboratory. Using samples provided to him by the two former University 1 professors in 2007, De Yoreo's analysis showed that the particles were mainly composed of carbon, which is exactly what we had found. The paper contained multiple experimental proofs that hexagonal particles made by the process of the science paper were unstable organic crystals of

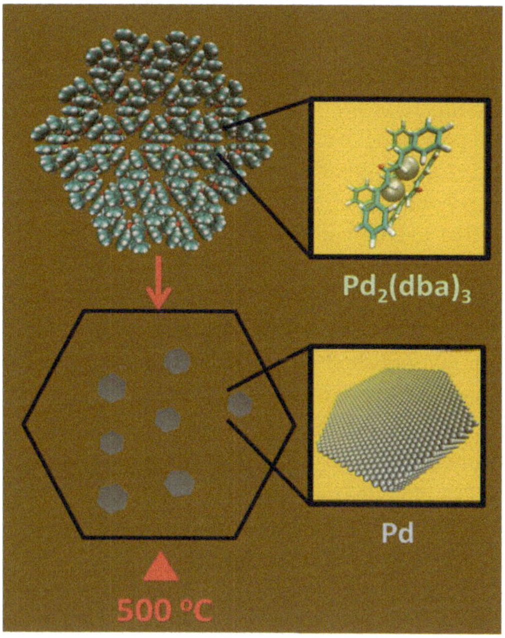

Fig. 1.3 Artistic representation of the form of the hexagonal platelets based on the available data from X-ray crystallography, electron diffraction, electron energy loss spectroscopy, and energy derivative spectroscopy. This illustration shows an idealized view of how the carbon portion of the hexagon may be vaporized at a temperature of 500 °C leaving only small Pd crystals behind

the palladium-containing precursor, but not palladium metal. The particles were so unstable that they degraded within a week at room temperature. They were melted under a desk lamp or dissolved in common solvents. A particularly telling experiment showed that the hexagonal particles pyrolyzed (vaporized) at a low temperature, indicating that they could not be palladium. Following heating to 500 °C, which is far below the melting point of palladium, the residue left behind consisted of tiny palladium particles consistent with a composition of ~5% palladium [24]. The tiny palladium particles were a residue from the small amount of palladium present in the palladium-containing precursor. The *Particle* article also showed a hexagon that had been grown to sufficient dimensions that it could be studied by X-ray crystallography. The crystalline material was of sufficient quality to prove that the structure of the molecules in the hexagon was composed of the palladium-containing precursor, not metallic palladium. This distinction is shown in Fig. 1.3. The top hexagonal structure is a cartoon of a hexagon composed of the palladium-containing starting material and the bottom one of palladium atoms. A hexagon made of palladium atoms may be of interest since it could have interesting properties for catalysis, but a hexagon composed of the palladium-containing starting material is a curiosity with no utility.

In the face of this overwhelming evidence, one might think that a retraction or at least a correction of the *Science* paper would be in order. After all these scientific publications were in the journals, in an interview in 2014, one of the former University 1 professors confidently claimed that his group had shown that the hexagons were composed of palladium [38]. They even received support from members of the University 2 administration who went on the record to say that they were confident that "research misconduct has not occurred" and suggesting that the matter should have been "settled in the journals". The statement made a university administrator ignored the fact that the matter had been settled in the journals. We had published overwhelming evidence showing that the Science paper was not only incorrect but misrepresented the facts [24–28, 39, 40, 46], and our findings had been corroborated by a second research group from a national laboratory [23]. This subjective judgement by an administrator from University 2 also ignored a federal report to Congress stating that the data in the *Science* paper were falsified. The final NSF reported called the objectivity of the process at University 2 into question. Why would a university support professors in the face of a federal report and substantial published scientific evidence that their publication is falsified? The federal report did not inform the public that University 2 had invested significant funds in the hexagon research. University 2 had an institutional conflict of interest that was never mentioned in anything written about this case, but it emerged behind the scenes in numerous ways. I will document this claim at the appropriate juncture in this book when discussing the use of intellectual property law and legal threats to intimidate anyone who would look into the veracity of the *Science* paper.

To understand how university financial interests control the course of research misconduct investigations we need to understand how science has evolved in the modern university. Universities have enticed professors to play a major role in the commercialization of technology with few checks and balances. Creativity in grant writing has replaced scientific creativity as the top criterion for success. Many scientists have come to believe that their role is so vital in supporting the university's mission and that the search for truth is ensconced in that mission statement that the "self-correcting" nature of science has eradicated ethical concerns. They are aided in this belief by university administrators who see only the good that comes from bringing in research funding and building a world-class institution. These good intentions ignore the ethical ramifications of the enormous financial pressures that faculty confront in the modern university.

Chapter 2
The Clash Between Scientific Skepticism and Ethics Regulations

> *"From time to time [...] one or another agency polls the American public concerning those communities within our society who are trusted or found credible – the Congress, physicians, bankers, newspapers and scientists. Invariably, scientists turn out to be close to the top of the chart of those viewed credible and trustworthy. Although there have been a few regrettable incidents, I continue to believe that confidence is well deserved."*
>
> *Philip Handler, Testimony in Hearings for the Subcommittee on Investigations and Oversight of the U.S. Congress, March 31, 1981*

There was a time when science enjoyed wide respect in society. The anti-science movement consisted of a small group of crackpots. Scientific credibility has decreased markedly since Dr. Handler testified before the U.S. Congress in the first hearings on issues of scientific ethics in 1981. Dr. Handler dismissed concerns about ethics only to have one of the greatest scandals, up to that time, erupt within 3 weeks of his testimony. I will discuss the Darsee case in some detail on the following pages. Traditionally, scientists were hard-nosed skeptics who needed to see multiple lines of evidence and every possibly control experiment to believe a result was reliable. Today, the public is skeptical and many doubt that scientists are honest. The reasons for this are corporate funding, the growing importance of real-world issues such as climate change, and threats to the ecosystem, but also scandals involving falsified data, companies based on bad science and the increasing corruption of a once clearly-defined mission to search for the truth. Following the Darsee case, reporters Broad and Wade wrote a sensationalized book about scientific integrity entitled "Betrayers of the Truth." [8] At the time, few scientists worried that this title or the trends described were a significant concern. Fast forward to today and we see that a large segment of society has lost respect for science. The fate of scientific funding will be decided at the ballot box. We could blame the evolution of political polarization in our society for the current danger. However, scientists who used to

S. Franzen, *University Responsibility for the Adjudication of Research Misconduct*, https://doi.org/10.1007/978-3-030-68063-3_2

seem beyond reproach are now seen by many as just another profession subject to corruption. I feel that we scientists must ask ourselves how we contributed to this predicament.

Scientists tend to be skeptics because their training teaches them the difficulty of predicting outcomes of experiments and experience shows them the difficulty of finding agreement between experiment and theory. The belief that empirical investigation is the only way to test a hypothesis is central to scientific skepticism. A wider view of science in a historical and sociological context informs us that the process of scientific discovery involves aspects of human intuition and imagination that transcend the traditional view of the scientific method. Thus, a second important aspect of scientific skepticism is reliance on common sense and instinct to reject hypotheses that do not pass a "smell test". In a more formal sense, ideas that fail to have a proper theoretical explanation or appear to violate established principles are often regarded as suspect. Harry Collins and Trevor Pinch have provided an excellent example the role of intuition as a guide for debunking poor science in their chapter on "Edible memory" in the book "The Golem: what everyone should know about science" [82]. Edible memory consisted of a hypothesis that an organism could obtain a memory of learned behavior by ingesting some part of a deceased organism that had acquired the memory. The idea sounds preposterous, but for many years several scientific research groups studying planarian worms, and even mice, promulgated the idea that memory could be passed on from a deceased organism to a living one individual by ingesting some part of the body of the deceased. The idea of edible memory was eventually discredited or perhaps, more accurately, simply pushed to the sidelines and forgotten. Collins and Pinch make the important point that many scientists refused to accept the statistical evidence provided in favor of the theory because there was no underlying theory or satisfactory explanation for the phenomenon. Eventually, when a sufficient number of research groups with good reputation could not replicate the findings, the notion of edible memory fell into disrepute. The demise of edible memory was not a negative experiment because it was impossible to design a negative experiment. There was a more direct way to dispense with the theory of edible memory. The fact that the theory could not be disproven, i.e. falsified in the meaning of Popper, means that it is bunk. This application of scientific skepticism is common sense.

A hypothesis should have a reasonable mechanism that explains the observations. Nonetheless, the criteria for a "reasonable mechanism" are subjective. Skepticism helped to debunk edible memory, but in the realm of ethics, skepticism may not always be a helpful impulse. Scientists may feel that their common and shared experience of a dedicated search for truth as part of the ethos of science makes it exceedingly unlikely that anyone could get away with falsification or fabrication. They tend to doubt that anyone would have motive and therefore there is no "reasonable mechanism" for falsification, so to speak. Even when confronted with evidence of misrepresentation they may seek another explanation and refuse to believe that any fellow scientist would attempt something so foolish. Skeptics who refuse to see the possibility of motive to commit falsification of research are confusing the integrity of science with integrity of individual scientists.

A person who knows how to manipulate the psychology of scientists can play on scientific skepticism to take advantage of the systems put in place to assure the quality of scientific publication and funding. Scientists will often point to the stringency of their demands for data, control experiments and other checks as evidence that the chances of getting away with fraud are vanishingly small. But the conviction that science somehow self-regulates, and the doubt that anyone would ever fake data, exposes scientists' gullibility [8]. This gullibility is evident in the naïve remarks by Phillip Handler in his 1982 testimony before the House Subcommittee on Investigations. In his capacity as President of the National Academy of Sciences he spoke for the scientific community when he insisted that the problems of falsification and fabrication were insignificant and managed well by checks and balances already in place. He gave his testimony only 3 weeks before one of the greatest frauds on science up that time was revealed. Starting with the first suspicion in October 1981, the John Darsee case rocked the scientific world and beyond. Darsee had risen to the highest levels in the field of cardiology at Harvard Medical School and gained the confidence of some of the leading medical professionals of his day before it was discovered that he had routinely falsified data or made up results out of thin air [8]. During the intervening decades there have been hundreds of frauds and falsifications and many thousands of retractions for various ethical reasons. The research misconduct legislation of 1985 spawned 20 years of effort that has resulted in new regulations in federal agencies. However, it is my contention that the regulations have not prevented falsification and fraud. In fact, the evidence present in this book suggests that the current implementation of federal regulations have aggravated the problem. Just to give one salient example, ironically a new scandal in cardiology at Harvard surfaced in October, 2018 when Dr. Piero Anversa was found to have falsified data in 31 publications over a period of more than 10 years on the use of stem cells to regenerate damaged heart muscle [83]. The Dean of Harvard Medical School stated to the New York Times reporter Gina Kolata, "The number of papers is extraordinary, I can't recall another case like this." [84] Institutional memory does not appear to extend to the Darsee case in 1981, which was also in the Cardiology Department at Harvard. This is not the only example where lightning struck twice. The now discredited argument that falsification is so rare that it is not a concern has been replaced by various arguments that reassure us that scientists are effectively identifying and eradicating fraud in science by a combination of rigorous peer review and the threat of research misconduct investigations by the federal agencies. Writing about the fact that Marc Hauser's extensive fraud (also at Harvard) was finally exposed after more than a decade of falsified reports, Art Markman wrote in Psychology Today, "There is no point in scientific misconduct; it is always found" [85]. But how is misconduct discovered? How long does it take and how many people are harmed in the process of finding it? Studies have exposed the much-talked-about "self-correcting science" as a myth [86, 87]. These studies show that few individual scientists are willing to correct their own mistakes when they are pointed out. The reality is that we have little data on ethical transgressions. Philip Handler in his testimony before Congress said "It [research misconduct] occurs in a system that operates in an effective, democratic and self-correcting mode … please understand, sir, that none of us has a real data base; none of us knows the real magnitude of the problem…" [88]. Herein is the issue; no one knows how large the problem is. Since it is hidden from view, Handler and many leaders of scientific institutions have subsequently appealed to the skepticism of their colleagues to cast doubt on the significance of the problem.

2.1 The Controversial Nature of Scientific Fraud

The term fraud is most frequently used to describe various schemes to achieve financial gain by identity theft, stock manipulation, or other misrepresentations. The term scientific fraud can be used in an analogous manner to mean gross misrepresentation of scientific facts, which may be for personal gain or fame. Most cases of research misconduct are less serious than fraud. Most falsifications are image manipulations or misleading statements, which may or may not change the interpretation and accuracy of a published report. It is possible to treat data or make observations in an ambiguous manner that would not rise to the level of falsification or fabrication and yet are dishonest. Sometimes scientists push past the ethical limit in their marketing of science in the effort to attract grant funding or investment. The scientific community has become tolerant of marketing in various forms and this alone has created an ethical dilemma. On the other hand, to claim that one has done an experiment, but in truth not to have done that work, to have no supporting data whatsoever, or cherry-picked data, is clearly a fabrication. In practice, such fabrications are essentially impossible to prove unless one has access to laboratory notebooks or original data. Unless an observer was in the laboratory and witnessed the discrepancies, the only other way that an outsider could discover a true fabrication would be expert use of statistics. If a co-worker expresses concern for either of these types of potentially unethical behavior, excessive exaggeration or cherry picking, the viewpoint of an outside observer will often be shaped by the respective status of the person making the allegation and the person whose research is being questioned. This social context of any such discussion will likely determine whether the target of the skepticism is the presentation of the research or the person raising the issue.

Scientists have a deep-seated concern that allegations against other scientists can lead to abuse. The use of the term scientific fraud as an allegation heightens this concern. It would be naïve not to consider the need to defend against detractors of science who doubt the validity of science in large part because they do not understand it. Opponents of research on climate change from outside the scientific community have alleged that the observation of the much-discussed global temperature rise has been based upon cherry-picking of the data. One famous example of this controversy created by allegations of wrongdoing is the science based on analysis of tree rings that led to the famous "hockey stick" graph showing the dramatic increase in temperatures in recent decades. In a prominent publication Dr. Michael Mann and collaborators showed that the temperature rise is a significant departure from the range of temperatures on earth during the past 400,000 years [89]. This study of temperature trends became a charged issue after the leaking of emails at East Anglia University by Russian hackers. One phrase from the private communications taken out of context made reference to "Mike's *Nature* trick" referring to a special treatment of tree ring records and other "proxy indicators" to examine temperature trends over the last six centuries [89]. The scientists had opted to leave out more recent tree ring data in the tree ring temperature correlation because of a well-known effect subsequent to 1950 that has led to decreases in the growth of trees in northern climates [90]. Climate bloggers and industry groups publicized the email breach and alleged manipulation of data. The issue rapidly escalated into government circles and was then seized on by the State Attorney General of the Commonwealth of Virginia and by members of Congress who named the scientists whom they considered should be investigated and ultimately punished. The furor over the emails triggered a research misconduct investigation by the NSF and seven other agencies, all of which exonerated the research scientists. The banter of the researchers about editors may have lacked professionalism, but poorly worded comments in private emails do not constitute a breach of ethics. The reason for the intense scrutiny was that the researchers had presented evidence for the theory that the global temperature rise in recent years is linked to the emission of greenhouse gases. The political nature of the claims of fraud can be seen in the fact that these were made by think tanks, public relations firms and others paid by oil companies and their investors [91]. Despite the exoneration by the federal agencies there has been continuing scrutiny of emails by these and other scientists by means of the FOIA mechanism. FOX News has dubbed the recent release of 5000 more emails from East Anglia University as "ClimateGate 2.0" [92]. The harassment is obvious at East Anglia, which keeps a public log of the large number of FOIA requests (and the UK equivalent) that it has answered. Such harassment of scientists has a damaging effect on the responsible conduct of science and flow of information to government and to the public. When such tactics bring powerful forces into the arena of research misconduct there is a chilling effect on specific environmental and health-related areas, science and on the adjudication of scientific ethics in general. It is also devastating for how scientists perceive the research misconduct system.

A memo dated June 28, 2011 from the Board of Directors of the American Association for the Advancement of Science strongly condemns attacks on scientists [93]. The memo also states "The scientific community takes seriously its responsibility for policing research misconduct, and extensive procedures exist to protect the rigor of the scientific method and to ensure the credibility of the research enterprise." It is indeed unfortunate that the "ClimateGate" emails became the basis for a major research misconduct investigation because this incident reinforces fears among scientists that the allegations of research misconduct will be used to harm scientists rather than to deter wrongdoing. The fear of abuse for political or commercial ends is one likely explanation for the weak wording of the regulation of research misconduct by federal agencies (OIGs) during the policy debate of the 1990s. In my opinion, scientists missed an opportunity to take charge of the ethical problems themselves when the regulations were being drafted. Scientist's skepticism regarding the potential dangers of federal regulation blinded them to the larger danger of putting their own universities in charge of the process of adjudication. The poor practice of giving universities the charge to investigate their own faculty is a central hypothesis for the failure to properly adjudicate research misconduct in the United States. However, we must bear in mind that scientists themselves advocated for this system. They wanted protection against frivolous or politically motivated allegations. They also believed that scientific ethics was not a serious problem, but merely a question of a few "bad apples" [94].

The "bad apple" theory suggests that fraud is an aberration, an infrequent case of greed or ambition gone wrong. By contrast, the evidence suggests a prevalence of exaggeration and misrepresentation caused be a variety of factors ranging from pressure to obtain funding to self-deception and even blind ambition. Although many cases of fraud have been hidden from public view, the available evidence shows hundreds of frauds in fields ranging from psychology to physics. For example, the case of Jan Hendrik Schön [9] showed the world that it was possible for a group of senior researchers to publish fraudulent science in the most respected journals over a period of years in the physical sciences. Dr. Schön published an impressive series of papers in the highest profile journals describing results that showed promise in the field of molecular electronics. At the height of the fraud Dr. Schön published papers in *Science* or *Nature* nearly every week for a year! Eventually, other researchers noticed a curious similarity in the noise in his experimental figures. Noise should be random, but the noise in two of the published experiments was identical, which is impossible to observe in actual data. It is sobering that none of the collaborators on the publications noticed the fabrications, but rather they were detected by someone outside the research group. Suppose Schön had been only slightly more sophisticated in his deception. If, instead of using the same artificial noise in different experiments, he would have generated noise using a random number generator so that each data set had different random (but falsified) noise, this innovation alone would have made it much more difficult to prove the case.

An examination of the trajectory of the Schön case suggests some disturbing trends that should give researchers cause for concern. The first implication is that peer review and editorial oversight are not working well. Schön published in the most prestigious journals with a frequency that broke all records. Although the editors in *Nature* and *Science* are looking for a story that will interest the readership, one would hope that the review process would detect obvious mistakes. However, editors also have an incentive to believe a radically new result, because it will propel their journal into the limelight. Vanity science magazines thrive on results that are based on risky premises, rather than solid science. Although the editors claim to have taken measures to strengthen peer review procedures in recent years, they have yet to show that they have prevented falsifications and fabrications from being published in the best scientific journals.

2.2 Failure of Referee Review of Journal Articles and Academic Self-policing

The reason that federal regulations arose in the first place is that scientists have not accepted the role of policing themselves. Given the way that peer review is *supposed* to work, it is hard to see how data are ever published that do not meet the highest ethical and scientific standard. A scientist may legitimately ask; how do other scientists publish outright falsifications when I have such a hard time convincing the reviewers that my carefully conducted science is correct? But neither refereed review of journals nor peer review of proposals is working the way it was designed. On the one hand, overly critical reviewers can reject good science to squelch competition. For example, Dr. Hermann Muller won the Nobel prize based on data that had hardly been peer reviewed [95]. He knew that he would have a difficult time proving that genetic mutations caused by X-rays where chemical rather than mechanical damage and that certain competitors were hostile to his ideas. Here is a case where intuition led to acceptance rather than rejection of a scientific hypothesis. On the other hand, the publish-or-perish mentality and growing numbers of scientists have led to an explosion of publications, some of them flawed or even falsified [96–98]. The explanation for failures is that there are lapses in review due to cronyism, laziness or outright fraud that permit misrepresentations to slip through. Although scientists may be skeptical based on their personal experiences, there is abundant evidence of fraud in the process of refereed journal review [99–101]. At the center of these extremes is a great moral debate. Some scientists feel that because they only know ethical scientists there is strong reason to doubt that unethical behavior is a significant problem. This leads to studies of ethics that justify the status quo. To make this argument one begins by debunking the idea that scientists are under greater pressure than they used to be. One study supporting this notion shows that first author publication rates have actually decreased somewhat over entire twentieth century [102]. It is argued that total publication rates have gone up because of increasing numbers of authors per article [102]. This finding has been interpreted to mean that recycling of data and publishing "minimal publishable units" are less serious problems than anticipated. However, this observation does not address either the quality of publications or the rejection rate. In fact, declining number of first author publications may be a sign of greater pressure since the expectation is and always has been that scientists will publish prodigiously. Since the number of scientists has increased it may still be the case that there is a greater absolute number of ethical issues in scientific publication. While meta-data analyses provide specific insights, one needs to examine science and scientific publication in a holistic manner. We must acknowledge, as Handler did, that we just do not have enough data to know whether fraud is increasing or decreasing. I would argue that this is part of the problem. The "data" are kept secret by way that research misconduct regulations are implemented. We are working in ignorance of the true extent of the problem. But every time we look for circumstantial evidence of ethical problems we find that examples abound.

Recently a series of major scandals in the peer review process in several journals has come to light, in which thousands of papers have been accepted based on falsified peer review [100, 103–105]. The first indication came from some suspiciously rapid reviews of the papers of a Korean scientist [106]. When confronted, the scientist readily admitted that he had written his own reviews. As this matter was investigated it was found that there were many such schemes [104, 107]. Several publishing houses reported that they had discovered that security flaws in their journal submission websites had been exploited [104]. Fake peer reviews were purchased from vendors according to a report from the Committee on Publication Ethics (COPE), "COPE has become aware of systematic, inappropriate attempts to manipulate the peer review processes of several journals across different publishers. These manipulations appear to have been orchestrated by third-party agencies offering services to authors." In addition, a plethora of on-line journals has provided a feeding frenzy of pay-to-publish services that have dubious peer review [96]. It is not only the for-profit internet journals that have issues. Retraction Watch has reported hundreds of cases in respected journals along with thousands of cases in the on-line journals. The peer review system harms good science as well. A recent study of the peer review process concludes that nearly half of all journals do not ask reviewers to address critical questions such as whether the data support the conclusions [108].

The failings of peer review alone cannot explain the epidemic of irreproducible results, cases of image manipulation and outright fraud that have been documented in recent years [109]. The fact remains that honest peer reviewers can fail to detect errors or misrepresentations. These are hard to detect particularly when authors have mastered the art of scientific marketing. Falsified results may sound reasonable and pass all standards of review if the reasoning is careful and there are no telltale clues of data manipulation. The accuracy of scientific work that fits this description would be difficult for anyone outside of the researcher's laboratory to discern. Reviewers can and should ask for more primary data in such cases. Any real breakthrough should stand up to scrutiny. We can already see mainstream journalism fighting for survival against the ruse of "fake news". The idea that there are different versions of the truth is anathema to journalism and science alike. When scientific reporting becomes subjective in the way current journalism has done, we have lost the essence of the scientific method. If that extreme prevails, then science itself is at risk of becoming a tool for technology development in the realm of political gain. Science itself will be lost.

2.3 The Advent of Federal Regulations and Their Reliance on Adjudication by Universities

Starting in the mid-1980s Congress initiated a process to implement federal misconduct regulations. By 2005 these had been set up in the federal Offices of Inspector General (OIGs) for each major agency that has a scientific research budget (NSF,

DoE, USDA, DoD, EPA etc.). The respective federal agencies defined their concern with the intent to falsify or plagiarize since these actions undermine the process of grant administration and defraud taxpayers. Consequently, the OIGs are law enforcement organizations. Since the agencies focus on the responsibility for appropriate use of funds, the regulations request that the recipient of those funds adjudicate any allegations. This is where a fundamental conflict of interest arises under current regulation. The regulation places the responsibility for the adjudication and correction of the research record in the hands of the "awardee institutions", most of which are universities. Universities are trusted stewards of the process and according to 45 CFR 689.4(a),

> "Awardee institutions bear primary responsibility for prevention and detection of research misconduct and for the inquiry, investigation, and adjudication of alleged research misconduct. **(a)** In most instances, NSF will rely on awardee institutions to promptly:
> Initiate an inquiry into any suspected or alleged research misconduct;
> Conduct a subsequent investigation, if warranted;
> Take action necessary to ensure the integrity of research, the rights and interests of research subjects and the public, and the observance of legal requirements or responsibilities; and
> Provide appropriate safeguards for subjects of allegations as well as informants.
> **(b)** If an institution wishes NSF to defer independent inquiry or investigation, it should:
> Complete any inquiry and decide whether an investigation is warranted within 90 days. If completion of an inquiry is delayed, but the institution wishes NSF deferral to continue, NSF may require submission of periodic status reports.
> Inform OIG immediately if an initial inquiry supports a formal investigation.
> Keep OIG informed during such an investigation.
> Complete any investigation and reach a disposition within 180 days. If completion of an investigation is delayed, but the institution wishes NSF deferral to continue, NSF may require submission of periodic status reports.
> Provide OIG with the final report from any investigation."

The NIH ORI and the other agency OIGs have similar definitions. In practice, NSF will almost always "defer" an independent investigation since there are so few resources in the NSF OIG (or any of the other OIGs) that they could not possibly conduct independent investigations of most cases. So great is the trust in universities, that little instruction to universities is provided in the regulation. The regulation fails to recognize that if a university takes the charge seriously, it will require a major input of time and effort for a process that risks a loss of prestige and funding. Since cases are confidential, university self-interest can derail an investigation in ways that are seldom documented. While administrators will justify confidentiality as a measure to protect the privacy of individuals, in practice it is used as a protective measure to shield universities from scrutiny.

If one assumes that the mission of universities is to promote the dissemination of knowledge to the public, it does seem natural that they should accept investigation of allegations of falsification or fabrication as a serious obligation. Unfortunately, this view of universities fails to take into account the dramatic change in university funding and the administrative culture that has accompanied the evolution of the modern university towards corporate status. In practice, the interpretation of OIG regulations requires that universities manage the inquiry and investigation entirely on their own. Federal agents refuse to interpret the regulation for fear of meddling.

The OIG answers to questions are vague because their assumption is that universities have an intrinsic commitment to the process with which the OIG should not interfere. Only at the end of the process when a university report is submitted to an OIG will the federal agency make any comment on the process. Unless there are serious irregularities, an OIG would normally approve the report and accept the university conclusion. If an OIG decides to conduct its own investigation, it will inform the university and usually that is the only contact with the university unless the OIG requests documents or wishes to make a site visit to interview witnesses. In such cases, the only contact between the OIG and university will be a formal close-out memorandum, which is filed on the OIG website in redacted form. The OIG will inform the respondents, the whistleblower and the university of the website address and number of the memorandum. If a finding of research misconduct is reached a letter of reprimand will be sent to the respondents and included in the report. Only if the university decides to take action based on the report would the rest of the world know that an investigation was concluded.

University administrators are caught between the need to comply with federal requirements to investigate and their own interests. Frequently, the university will do the minimum required to comply with the federal agency and justify this conduct in terms of a cautionary approach, not wanting to overstep the bounds set by the agency. Administrators tend to worry there may conceivably be a negative repercussion taking an action that the OIG would deem inappropriate. For this reason, a university can always justify inaction by claiming that they have received no explicit instructions to act, e.g. retract falsified publications or reprimand authors. However, university administrations may not recognize that *a federal agency OIG will never instruct a university to take any action* beyond two general instructions, first to conduct an inquiry when an allegation is received and second to submit a report at the conclusion.

Each federal agency has written its own separate research misconduct regulation specific to the OIG for that agency. Given its central role in the funding of science, one can consider the NSF research misconduct regulation in Section 45 of the Code of Federal Regulations (CFR) 689.1 *et seq.* as the prototype for the federal research misconduct regulations. In order to reach a finding of research misconduct, the regulations first require that the university find falsification, fabrication or plagiarism was committed and second that one of those acts was a significant departure from the norms of the academic community. The third and most challenging requirement of the regulations is that the act must be proven to have been reckless or intentional. This three-part definition of research misconduct common to all agencies is clearly an attempt to prevent frivolous or politically-motivated allegations [110]. But the wording used to protect scientists from unfair investigations results in ambiguity in the regulation. NSF regulation 45 CFR § 689.2 states that "*Research misconduct* means fabrication, falsification, or plagiarism in proposing or performing research funded by NSF", but also requires establishing intent or recklessness. What happens in the case where a committee identifies a falsification, but fails to agree on whether there was intent? Such a committee conclusion was the reason that the NSF OIG did not accept university report in the hexagon case. But given the limited resources of the NSF OIG it may not be able to conduct its own investigation in every such case.

A common feature of OIG regulations is that any breach of ethical conduct is under the jurisdiction of a federal agency only if funding has been received from that agency. Although the language quoted from 45 CFR § 689.2 reveals a potential pitfall in the regulation it turns out that this language can lead to obfuscation in the hands of unscrupulous scientists and university administrators. Suppose a paper alleged to contain a falsification used NSF funding, but fails to cite it. Then technically the publication is not under NSF jurisdiction. It is up to the discretion of the university administrator in charge to decide whether to attempt to establish that there is NSF jurisdiction. However, the university administrator has a conflict of interest since establishing NSF jurisdiction creates a great deal more work and the risk for sanctions against the university. In fact, the administrator has a much greater incentive to establish that the NSF does *not* have jurisdiction and there is at least one example where a university unilaterally declared that NSF does not have jurisdiction despite evidence of NSF funding [111]. The NSF OIG deferred to the university judgment.

The regulation and the management by the OIGs both take great care to preserve the independence of the university. Therefore, universities are free to interpret the confidentiality clauses in the regulation in a manner that is even more restrictive than the agency. On the other hand, unless the university has posted a regulation concerning confidentiality it implicitly uses that of the federal agency. NSF relies on institutions to "Provide appropriate safeguards for subjects of allegations as well as informants" in 45 CFR § 689.4(a)(4) and in § 689.5 it is stated that "The identity of informants who wish to remain anonymous will be kept confidential". Here too, there is no guidance in how to implement this regulation to maintain confidentiality. Confidentiality prohibits anyone from knowing how many allegations are accepted or alternatively how many end up in the trash bin. We do know that the number of documented research misconduct cases is vanishingly small relative to the number of retractions. Most of the cases actually pursued involve students or post-docs. But when allegations are brought forward against a faculty member, the risk for the bottom line of the university creates an obvious institutional conflict of interest, which interferes with objective evaluation of an allegation. Faculty represent an investment by the university, and they may have grants that bring in overhead or start-up companies with a university equity stake [112]. This is the heart of the problem for research ethics in the modern university. In public universities, the administration has been dealing with budget cuts and a climate of great economic pressure. To conduct a research misconduct investigation that may result in harm to one of the "producers" for the university is extremely damaging to the institution. The pressure to downplay allegations of faculty research misconduct is enormous.

2.4 The Role of the Research Integrity Officer (RIO)

The role of determining if allegations warrant inquiry has devolved to the institutional Research Integrity Officer (RIO). The RIO is the gatekeeper who decides whether a case should go forward and plays a crucial role in the conduct of both inquiry and investigation. Given the crucial importance of the RIO it is remarkable that the role of the RIO is not mentioned in federal research misconduct policies [113]. One would hope that the RIO would reduce the number of cases through education and outreach to foster an environment that discourages misconduct. In cases of authorship disputes or some forms of plagiarism resolution can sometimes be achieved without resorting to filing of allegations [114]. Certainly, the use of a Faculty Ombud or approaches that can mediate a dispute or convince an author to retract a flawed paper are desirable and positive developments. But one must suspect that there are universities where informants (whistleblowers) will be discouraged from filing an allegation or that an allegation may be deemed frivolous because it involves university-owned technology or major grant funding. One can imagine that this would be done with some skill so that informants are not alerted to the reasons that their allegation does not warrant an inquiry. Alternatively, committee selection or the way in which the RIO presents the allegation to the committee in the inquiry phase may affect the outcome. In cases involving senior faculty, the RIO will normally be familiar with the relevant technology at the administrative level from a knowledge of the grant funding and patent portfolio of the faculty member. Therefore, the RIO will know when intellectual property (IP) is of value to the university. The RIO is also aware of how much the investigation can cost. The RIO will be under pressure to deal expediently with problems; adjudicate those that are obvious, decline those that have an element of revenge or animus, and manage the difficult ones to prevent them from escalating into an embarrassment for the university. Above all, the RIO is to maintain confidentiality. Because of the confidentiality of research misconduct cases, there is no way to know how many cases are rejected due to university conflict of interest. Sadly, this is a major reason that Handler's statement is correct that we have no idea how great the problem really is.

2.5 Skepticism Regarding Allegations of Falsification

Scientists tend to be highly skeptical of claims that data have been falsified by a fellow scientist. Of course, scientists are not supposed to know when allegations have been made, but in any serious case of a flawed publication it is difficult to keep anything a secret. The skepticism of the validity of an allegation is intertwined with the way that the regulation has made research misconduct a serious offense consisting of *intentional* falsification or fabrication. Because of the federal regulation, anyone who makes an allegation should provide proof not only of the lack of validity of the result due to misreporting, but also of intent to misrepresent the research. How many whistleblowers know this when they confront an ethical problem in research? Moreover, unless the evidence includes image manipulation, evidence of intent to falsify is usually difficult to produce. Image manipulation has certain similarity to plagiarism in that the evidence is self-contained in an electronic document [109]. However, evidence may be in a laboratory notebook or instrumental log files that are not accessible to an outsider. Even a witness or close collaborator may not have access to the data needed to prove that there has been a misrepresentation. Obtaining laboratory notebooks often is impossible without first starting an inquiry. Hence, a person who witnesses apparent discrepancies between methods used in the laboratory and claims made in a publication is in a difficult position. The only permissible charge under the regulation is one of "research misconduct". However, if the process does not prove *intent* to falsify or fabricate, the research misconduct charge will be dropped, and the respondents will be exonerated. Falsification without research misconduct is not actionable by the federal agency under the current implementation of the federal regulations. All this may sound unnecessarily legalistic for a scientific misrepresentation in a journal article, but this is how the regulations have been written and therefore how the adjudication proceeds. **The complexity of proving intent is a significant reason that the current system fails to deal effectively with falsification and fabrication.**

The current research misconduct system removes the necessity of a scientific evaluation of the facts. The federal regulation defining research misconduct does not rely primarily on scientific reasoning, but rather on evidence of the intent to falsify. On the contrary, scientists are likely to be less worried about intent than the fact that falsification of data can lead to a great waste of time and effort. If the investigative process focused entirely on whether the scientific facts supported the claims in a journal article, the issue of protecting confidentiality would not be necessary. It would not even be possible since the data and publication are both in the public domain. If in the course of such scrutiny evidence of data manipulation came to light, this would be grounds for retraction of a paper, regardless of intent. This discussion clarifies that confidentiality is contradictory to the culture of open exchange and debate and therefore it impedes any evaluation of the scientific facts by a committee. It results in sequestration of the data from the scientific experts who are likely to be in the best position to evaluate the data. If confidentiality were maintained perfectly, only the informant, respondent, RIO, university general counsel and committee would know of the case. But there are always leaks. When colleagues hear about a case but cannot examine any data because it is sequestered by the university, scientific skepticism leads them to doubt that the allegation is substantial. There is a strong tendency to give the accused the benefit of the doubt and to fear the possibility that an allegation is an attack by a vigilante or based on a personal grudge. Many scientists are aware of their own transgressions, large or small. Since scientists will admit in private that they have skeletons in their closet, they have an incentive to distrust ethics investigations. Thus, a whistleblower is in a vulnerable position and surrounded by a great deal of skepticism.

2.6 Skepticism of a Motive for Falsification and the Suspicion of a Motive for Revenge

Since scientists pride themselves on openness, transparency and honesty, the report that a fellow scientist has fabricated or falsified data will generally be greeted with suspicion. This suspicion is compounded by fear that the federal regulation may interfere with the goals of science. Many of these attitudes are fueled by a natural antipathy of most scientists towards accusations or investigations, which leads them to know as little about the regulations as possible. Ironically, federal research misconduct regulations were written with the expectation that scientists would give selflessly of their time and assist in the adjudication of research misconduct in an objective fashion. While any scientist will most certainly agree that it is important to maintain scientific integrity, scientists feel uncomfortable in the extreme in being placed in a position where they must judge someone else's ethics. To judge their competence, the merits of their argument, the quality of the data, or the soundness of their conclusions are all part of the job of an academic. But no scientist wants to judge someone else's integrity.

These issues complicate the position of a scientist who has been cajoled into serving on a research misconduct committee. The sequestration of data and lack of communication in a research misconduct investigation are completely foreign to the experience of a research scientist. These factors discourage scientists from agreeing to sit on an investigation committee and can negatively affect how they respond to the issues put before them. Those not serving on the investigating committee often know little about the regulation and will have no appreciation for the constraints that limit the disclosure of the facts. The skepticism of scientists demands a transparent process, but the regulation ensures the opposite.

When patented technologies are involved, there can be a financial motive for falsification. When money is the motive and lawyers are called in to threaten scientists who observe questionable practices in the laboratory, the balance of the motives may be quite different than the quaint view of squabbling professors bickering over some arcane scientific issue. Under current regulation a lawyer's threats are kept secret by a confidentiality requirement. Because of the requirement for confidentiality, most observers have no idea if money is involved or threats of legal action have been made against the university. The informant or even the members of an inquiry committee looking into an allegation may subjected to threats with legal action. Meanwhile, university administrators may have a different set of concerns. Protection of the university's reputation is a focus since investors will view the credibility of the TTO and institution as a whole as part of their investment in early stage spin-off companies [115].

Much of the discussion of ethics in the scientific community is based on skepticism of scientists who hear about an allegation of falsification and believe that it must be a dispute because have a difficult time believing that another scientist would falsify data. Skepticism is part of a scientist's training, but it also may lead certain scientists to a refusal to consider the possibility that a colleague could have falsified data. The refusal to consider the possibility of falsification makes scientists more gullible than the person on the street who hears about such a case [116]. If scientists pretend that ethical issues are absent from scientific investigation they run the risk of fooling themselves or being duped by others.

The book "Ending the Mendel-Fischer Controversy", by Fairbanks and Franklin provides a skeptics view of the suspicions of the statistician Ronald Fisher that Gregor Mendel's data were manipulated in some way since it is highly improbable to obtain such an ideal distribution of data among the various gene combinations [117]. Although Mendel is credited with laying the foundation for modern genetics based on the observation of traits of pea plants, Fisher noted that Mendel's statistics are "too good" for the simple model of dominant and recessive genes [118]. Irrespective of exceedingly low probability ($< 0.007\%$) that Mendel's numbers could have been observed as reported, Fairbanks and Franklin exonerated Mendel posthumously [117]. In the final analysis, Fairbanks and Franklin cannot prove that Mendel is innocent of the intent to falsify any more than Fischer could have proven that Mendel intentionally manipulated the numbers. Why is it important to establish whether Mendel committed research misconduct by today's standard? Would it change our regard of Mendel as the founder of modern Genetics? Blind faith in scientists as ethical human beings is as harmful as uncritical trust in universities arbiters of scientific misconduct. Gregor Mendel may well have been at the same time an Augustinian monk, a great scientist and someone who manipulated the numbers, perhaps without realizing the significance of his actions.

In the modern era, federal research misconduct regulations give universities the responsibility for adjudication of allegations of fabrication and falsification. Dr. Franklin as the Head of the Standing Committee on Ethics at University 2 was given the charge to look at the evidence in the hexagon case. However, according to the National Science Foundation final report the University 2 committee report did not examine the available evidence and was "not accurate" [42]. Unlike the case of Mendel, in the hexagon case there was ample evidence of falsification and there were witnesses. But Dr. Franklin's committee did not examine the evidence or interview any of the witnesses when he exonerated the two professors recently hired by his university by declaring that the allegations were "without foundation". To say that someone has made a baseless allegation is itself an allegation against the whistleblower. It took 7 years after those events at University 2 for the final report from the Federal agency to vindicate the whistleblower. The conferral of power to universities is implicitly based on the assumption that universities are ethical. The federal regulation does not contemplate the harm that can arise if a university fails to live up to its ethical responsibility.

The combination of conflicts in regulation and external funding pressure has caused a crisis in scientific ethics in the modern university system. When funding itself becomes a dominant metric for scientific achievement, the incentive to exaggerate or deceive oneself increases, sometimes merely as a matter of survival, but also in the competition to gain attention for one's research. The corresponding focus on the financial aspect of misconduct has led to an all-or-nothing view of ethics. Under the current system, the conclusion of an investigation will only be acknowledged by the agency if a scientist has committed research misconduct. In that case, a redacted announcement is made on an anonymous website managed by the NSF. However, in the absence of proof intent or recklessness, the NSF misconduct regulation results in exoneration of the respondent. It is difficult to imagine that anyone intended to create a system that permits the exoneration of falsification and fabrication, as our current regulation does. I believe that the thought process that leads to the defense of Gregor Mendel's manipulations of data stems from how our funding agencies and universities would treat such a case today. By the standard of federal regulations today, Mendel did not commit "research misconduct" because he did not falsify "recklessly" or with "intent", despite the strong evidence that he falsified data, as originally proven by Fisher based on statistical grounds. The notion that Mendel should be posthumously exonerated for not understanding the procedures of modern science is a subjective value judgement that should not intrude on the serious problem facing science today. The scientist-entrepreneurs of today are not monks as Gregor Mendel was.

Chapter 3
Scientific Discoveries: Real and Imagined

For the ideologists of science, fraud is taboo, a scandal whose significance must be ritually denied on every occasion. For those who see science as a human endeavor to make sense of the world, fraud is merely evidence that science flies on the wings of rhetoric as well as reason.

–William Broad and Nicholas Wade, Betrayers of the Truth, 1983

Science is a boring and frustrating occupation. And yet, it is the most fascinating subject in the world to those who master it. The daily humdrum of experimentation and record keeping can be mind-numbing. Nonetheless, at the end of a long period of experimentation and observation a most elegant world view emerges, and those who have patiently slogged through a mass of data may find an explanation for a natural phenomenon. Hypotheses can build on previous experiments to provide verifiable predictions. Those who understand the interplay of experiment and theory cannot help but be awed by the beauty of science. To the citizenry on the outside of this awesome process, science and technology is sold as a contribution to the common welfare and prosperity, our key to mastering agriculture, energy, medicine, transportation among other aspects of society. The abstract search for truth, coupled with benevolent expectations of others, may explain why scientists are not predisposed to recognize the possibility of falsification and fabrication of scientific data by their colleagues. The messiness of human psychology is not written into the idealized dialogues with Nature and society at large, with which scientists are engaged.

The path to scientific mastery typically requires a quantitative evaluation of repetitive procedures. Science is not usually portrayed in this way. Archimedes jumped out the bath once and shouted "Eureka!" after arriving at the insight that he displaced his weight of water in the bathtub. Presumably, he did not jump out of the bathtub on a regular basis only to enumerate the days he felt buoyant against those he did not. We believe that it was a sudden insight that led to Archimedes principle. Yet, the validation of those insights is fraught with difficulty. We may have a keen

S. Franzen, *University Responsibility for the Adjudication of Research Misconduct*, https://doi.org/10.1007/978-3-030-68063-3_3

intuition or even a well-developed theory, but the verification process is long and arduous. To the layperson, it may be surprising to learn that first experiments are rarely conclusive. The process of making a measurement can be enormously frustrating. Since mistakes in the laboratory are common, the process of investigation is intended to reveal mistakes by means of control experiments and consistency checks. Often, these additional experiments reveal inconsistencies in the results. While an unexpected result may be the first clue to a discovery, more frequently, a failed experiment is simply a dead end. Worse still, poor experimental design can result in the failure of the control experiments in ways that escape detection. Distinguishing the difference between an unexpected insight from an error in experimental design is crucial to good science.

Rewards in science follow priority. The first to discover or describe receives that professional credit. Publications are dated to record the timeline of discoveries for the benefit of the scientific community. To be first, however, a scientist must be fast. And since scientists build on the work of one another, the speediest *modus operandi* involves trusting the work of colleagues. There is no reward for checking the work of other scientists. Funding is granted to new ideas, new shoots on the tree of science. Watering the roots and picking off pests maybe essential to the health of the tree, but these tasks are not glamorous. Careers are made by being first. Being correct is left to posterity.

As the canopy of the tree of science proliferates at an accelerating pace, it becomes harder and harder to recognize whether a new shoot is new and has not been expressed in the same way by another branch. It is a bit like social media. How does one make oneself noticed when content is semi-infinite? To ensure that rewards will derive from being seen as first, there is a growing tendency among scientists to rely on jargony neologisms, flashy figures, and bragging rights to journal covers that have nothing to do with the science itself. Modesty is becoming obsolete or even a liability. Marketing is part of science today whether we like it or not.

With all the pressure to succeed in a system that rewards those who have a rapid succession of new results, there are scientists who succumb to the temptation to cut corners and publish results that are not properly supported by data. Sometimes this occurs because of great haste and misunderstandings in a research group. Other times it occurs because of self-deception and a refusal to take the time to look at all the possible angles. Regardless of how falsified data make their way into scientific publications, they cause an enormous waste of time and effort by researchers who follow on and try to use those results for a next step in scientific progress. There is a prevailing belief in the scientific community that any fraud will be quickly exposed by its lack of reproducibility and inconsistency with other published results. However, a survey of the history of science reveals that there has been a persistent undercurrent of ethical lapses. There has not been a history of rapid recognition of these flaws and even serious misrepresentations have persisted for decades in the literature.

3.1 The Evolution of Science and the Science of Evolution

The scientific method is based on the idea that there is an explanation, a hypothesis that will be tested by experiments and then either disproven or confirmed. Competing hypotheses naturally arise when disparate investigators try to reckon a group of imperfect measurements or observations. One hypothesis typically survives additional data and observations while the remainder are discarded. The selection process involves comparison of the theory with data and the survival of a hypothesis is based on the quantitative ability of the theory to correlate existing data and predict future outcomes. Nevertheless, even successful theories are not immutable. They evolve like a species in biology. Successful scientific theories may require modification to embrace new content. Biological evolution, on the other hand, is the result of a great number of random mutations from generation to generation that permit adapted individuals to survive and reproduce.

Biological evolution is a firmly established scientific theory that has had to endure attacks from outside by those who interpret it as a threat to their religious beliefs. At the other extreme, it has been abused by others who have used it to justify eugenics. Regardless of ideology, we can examine a brief history of the theory of evolution to understand the ethical dimension of science. Evolution is important to the hexagon case because the claim of an "evolutionary" technology, albeit in a test tube, was one aspect of the salesmanship used to publish a completely erroneous result. The central role of the theory of evolution in biomedical research permits us to compare its application to various disciplines. For example, while it is possible to conduct experiments on bacteria or fruit flies that clearly show the connection between genetic mutation and changes in morphology on a time scale of days or weeks that lead to a competitive advantage, the time scale for human evolution is tens of thousands of years at a minimum and the issues involving survival of the fittest are enormously more complicated. For this reason, the most direct confirmation of the theory comes from organisms that mature quickly, viruses, bacteria, fruit flies and so on. The evolution of bacteria can take place in an afternoon and is the basis of a test for mutagenesis known as the Ames test. Bacteria are exposed to a molecule for testing as a mutagen (DNA modifying agent) and thus likely carcinogen (cancer-causing agent). The bacteria are placed on an agar (food) plate with a minimal amount of an essential amino acid. If their mutation rate is high (i.e. if the agent being tested is likely a carcinogen) they rapidly change their ability to produce the missing amino acid (over several generations) and begin to grow on the agar plate. The control group, which has not been exposed to the carcinogen has a lower mutation rate and does not grow well on agar lacking the essential amino acid. In another example, the theory of evolution also explains antibiotic resistance in bacteria, which have evolved after contact with an antibiotic and mutated so that a small segment of the population can survive. These survivors then become the resistant population. Evolution also explains why cancer tumor cells change with time to evade toxic drugs. After being challenged with chemotherapy drugs some of the tumor cells mutate and survive, thus becoming drug resistant. These are but a few examples of the ways in which modern medicine relies on the theory of evolution for the treatments of many diseases.

Despite our sense that humans are evolving in response to environmental and social pressures we have no means to determine whether human biological evolution occurs in a time as short as human history. Science and technology have become intertwined with economic progress, health, diet, and security to the point that we may ask how science is affecting human evolution. While technology has improved life in many ways it has also brought us the threat of environmental devastation and nuclear war. When viewed in these terms the ethics of science is more than merely an academic exercise in rooting out plagiarism or retracting falsified journal articles. The ethics of science is at the heart of how science will be implemented as technology, how it will be taught and whether it will be used wisely by those who have power in our political system. To those scientists who fear that discussing this topic opens Pandora's box to the entire panoply of public criticisms of science, I must say that the box is already open. To ignore the ethical issues in science increases the vulnerability of science to attack when the scientific community least expects it.

The idea that scientific theories are progressing towards truth is deeply rooted in scientists' view of their work, but history shows that there have been many wrong turns and, at times, even regression of scientific understanding. There is also ample evidence that scientists are quite reluctant to accept new hypotheses when they conflict with the scientific ideas that have sustained them in their careers [119]. Incorrect theories, errors and even falsifications have persisted for decades in some cases, belying the myth that scientists will always quickly embrace new evidence that disproves a hypothesis. It would be comforting to think that ethical transgressions in science will be corrected by a process of survival of the fittest theory that parallels the analogous biological concept. The notion that science is self-regulating [120] or self-correcting [85] assumes that there are incentives for checking science for reproducibility. However, the economic pressures in modern universities do not provide sufficient incentives for studying reproducibility. Before considering how the recent history of the university system has affected scientific ethics, we should examine some examples of scientific ethical issues that have arisen in the context of a firmly established theory.

One interesting product of the natural selection of scientific theories is the theory of natural selection itself. Charles Darwin had to fight an uphill battle for acceptance of his theory against those who were discomfited by it on religious grounds, but he also had to compete with other scientists to receive the credit for the discovery. The empirical nature of discovery is such that it often occurs simultaneously in several different laboratories. If one scientist does not realize the next step, the discovery or procedure, another scientist will surely do so in short order. The incentive to be first can have an enormous influence on the temperament, and the ethics, of a scientist. Although Darwin received the credit for posterity, Alfred Russel Wallace reported his conclusions first and Darwin's friends had to intervene to ensure that Darwin would receive credit. Moreover, Loren Eisley has argued that Darwin got a significant intellectual boost from two articles by zoologist Edward Blyth written in 1835 and 1837, which Darwin never acknowledged. Darwin's contemporaries chastised him for not giving more credit to others, but Darwin chose not to share the limelight [121]. This type of simultaneous discovery and struggle for credit is ubiquitous in science. It can lead to pressures to take unethical steps in the conduct of the experiments such as rushing to be first to obtain a result, while becoming sloppy and careless in the process. This is such a common problem that one could write a book just about this aspect of scientific ethics [122]. It can also lead to plagiarism or denial of credit by not citing others who are either less well-known, perhaps in another country or otherwise unable to assert their right to acknowledgement for their contribution. Plagiarists who have stolen credit for other's work include Ptolemy who copied his theory from the a anonymous Greek astronomer on Crete in 150 A.D. to Americans, Europeans and Israelis who copied articles wholesale from the Soviet literature and many more recent examples [123–126].

A second ethical dimension arises in the honesty and accuracy in the recording of scientific data. A fundamental understanding of genetics was needed prior to general acceptance of the theory of evolution. We return briefly to Gregor Mendel's notebook where he explained his observations of inherited traits of pea plants using a theory of heredity based on genes. This work was done between 1856 and 1863, well before there was any chemical information to describe what a gene actually was [127]. Even though that Mendel's theory fit well with Darwin's observations, the scientific community (including Darwin) did not accept the validity of Mendel's hypotheses during his lifetime [128]. In 1882, Hugo de Vries used Mendel's concepts to explain genetic data relevant to the theory of natural selection [127]. Subsequently, modern genetics was put onto a firm footing by the statistical analysis of Sir Ronald Fisher in the early twentieth century. Fisher was troubled by his own conclusion based on a statistical analysis that Mendel may have manipulated data. The values Mendel reported were essentially perfect. Moreover, Mendel's values disagreed both with the notion of random error and also with known genetic factors based on subsequent observations of pea plant reproduction. Rather than believe that Mendel could have done anything unethical, Fisher suggested that perhaps a scribe played a role in writing down the statistically improbable results. While it is natural for scientists to want to defend each other, if we cannot recognize falsifications that occurred more than 100 years ago, we run the risk of denying those that happen today.

A third dimension of scientific ethics concerns wholesale fraud. While this aspect is regarded as a complete aberration, there are well-documented individual examples in an almost continuous stream throughout the history of science. One of the most famous frauds in the history of science emerged shortly after Fisher's work codified genetics in its modern statistical form. The gene theory of heredity was challenged Dr. Trofim Lysenko in the 1930s in the Soviet Union [129]. Lysenko's views on evolution were a throwback to LaMarckian evolution, which had many adherents in the Communist Party because of the attractiveness of the idea of behavioral heredity in creating a better worker. Lysenkoism became so powerful that its opponents were legally forbidden from publishing alternative theories in the Soviet Union in 1948 and the tyranny continued until 1964, 10 years after Stalin's death. Lysenkoism has been seen as a warning of the unfortunate consequences when nonscientists enter into scientific controversies with political motivations. There are other related examples where prejudice took precedence over objective observation. Studies of skull size as a measure of human intelligence, interpretations of IQ test and so on, for nearly 100 years had elements of this type of pseudo-science. In all these cases, there was ample falsification, which was often quite blatant. Yet, the scientists who carried out these studies, such as Dr., Samuel G. Morton, Alfred Binet and Cyril Burt, were respected members of the scientific communities of the United States and Western Europe. It is crucial to recognize that, for many decades, the scientific community accepted those thoroughly discredited ideas [8]. We need to consider the role played by logical reasoning, data collection and statistics in each branch of science to better understand how these ethical issues relate science as a whole.

3.2 A Hierarchy of Methods: The Role of Proofs, Logic and Statistics in Various Fields

The relationship between data and hypotheses can be described in a hierarchy of scientific methods based on their standards of proof`, the way evidence is obtained, and the role played by statistics. Mathematics is based on proofs. There can be a debate about what may or may not be proven in the future, but proven theorems are not debated. There may be ethical concerns about who gets credit for mathematical insights, but the logic of mathematical proofs precludes the kind of falsification that can be found in data-driven sciences. Physics, physical and analytical chemistry are applications of mathematics to phenomena in the natural world. Unlike mathematics the hypotheses are driven by data. There is room for debate about how to apply equations to natural observations as well as the relative magnitude of competing effects. As soon as data enter into science the issues of honesty and accuracy of data collection become ethical concerns. On the other hand, the established theories, embodied as the laws of physics, are so fundamental that we cannot imagine replacing them. However, they may not be complete. As one considers chemistry and the molecular sciences of synthetic chemistry, biochemistry and molecular biology one enters a realm where there is a great deal of debate over how processes occur. The analytical methods for resolving these debates are quite sophisticated. There is sufficient agreement to codify chemical knowledge in sets of rules and trends that have great predictive power because the statistics of chemical systems deals with large numbers. These sciences are based on hypotheses and certain of the established principles are called laws, but proofs do not have a direct application to these sciences. The rigor of analytical methods makes fraud difficult in such technical fields, but as these fields have grown more complex ethical problems have surfaced. Indeed, the hexagon case is one such example.

There is an increase in complexity of experimental observations and decrease in sample size encountered in biology, ecology and medicine. Sample size is a crucial aspect of how accurately a particular quantity can be measured. An analytical or physical measurement in chemistry involves hundreds of billions molecules at a minimum and can easily be a trillion times larger for some measurements. Even with such large numbers, precision of 1% is considered good in reported value. If we were to accept a single measurement as valid, we might consider the precision to be even higher, but by convention a minimum of three individual measurements is required. Then an average and standard deviation is calculated for these to give the reported precision of the number. In any data-based scientific field statistics is crucial for quantifying results and it is central to all data interpretation. The precision of measurement inherently increases proportional to the sample size. Moreover, as one moves from chemical to biological and geological sciences there are fewer observations possible, in part because the time scales are longer in life and earth sciences. There is also greater ambiguity in experimental design because there is a larger number of variables in the system. While variables in chemical systems can be described with rigor using thermodynamics and quantum mechanics where a well-defined set of variables describe the system, it is often not possible to be sure that all variables have been accounted for in biology, ecology and such disciplines. There is a corresponding difficulty of constructing control experiments as the number of observables increases and the size of data sets decreases.

As we consider any topic that involves human health the number of subjects in a study is often quite limited and the number of environmental, genetic and physiological factors that one must consider is potentially very large. The replication of experiments is difficult because of the constraints on samples sizes. While a biologist studying fruit flies may have thousands of specimens, clinical trials may be limited to a small group of patients who have a particular condition. Biomedicine and environmental science are also more prone to bias because the topics under study have direct impact on the quality of life of people, on companies and their profits and on the protection of the environment. The number of potential ethical issues increases dramatically in these fields.

Social sciences and economics are quantifiable disciplines, but our ability to conduct an experiment with a control group becomes more limited as the object of study becomes larger and more complex. For example, we cannot study the effect of education on the economy of Nigeria directly by comparison to a control since there is no country like Nigeria that is the same in all respects except its education system. In such fields of study, the observations are long-term observations of human activity subject to certain aggregate observables with as much quantitative rigor as possible applied to derive useful theories. Psychology is a discipline in which one approaches a borderline where it may be difficult to define a reproducible experiment because of the vast variations in humans, in addition to limitations in sample size and the length of time required to make a reliable set of observations. The large number of variables and subtle differences in personalities make it difficult to define control groups, i.e. groups who are observed under conditions that are identical to the study group except for one variable. Ethical problems in research are indeed more common in fields dominated by statistics, but the surprising observation is that serious falsification has been observed across the entire spectrum of scientific inquiry.

3.3 The Role of Statistics in Evaluation of Scientific Data and Hypotheses

In many branches of science, statistical analysis is applied to evaluate whether conclusions deduced from a subpopulation reflect the whole. Our example for this discussion will be a group of twenty individuals in a population who have a fever with a mean temperature of 102 °F. We will give a drug called Feverbreak to a sample group of twenty individuals and take their temperature 2 h later to determine whether there has been any change in body temperature. We will determine whether the new mean temperature differs significantly from the original temperature of 102 °F. We will then consider what this says about the efficacy of Feverbreak.

Real data burdened with errors typically have the shape of a so-called normal distribution [130] shown in Fig. 3.1, where measurement near the mean is more frequent than measurements at other values. The normal distribution is a symmetrical bell-shaped curve that has a total area of one. The fractional area within specified limits can be interpreted as a probability. In our example, the "null" hypothesis presumes that Feverbreak will do nothing to reduce the fever of the group with a mean temperature of 102 °F. The word "null hypothesis" simply means starting or initial hypothesis. If we consider a common application in medicine and the distribution of the null hypothesis has some symptom or disease, then scientists are looking for a change from the original distribution as an indication that a drug is making difference. Based on the data shown in Table 3.1 we can determine the mean and the standard deviation, σ, of the distribution of the individuals who have a fever. The standard deviation obtained from Table 3.1 is $\sigma = \pm 1^oF$ and it is related to the width of the normal distribution as shown in Fig. 3.1. Two hours after treating with Feverbreak, the distribution has a new mean, which is lower than the original mean

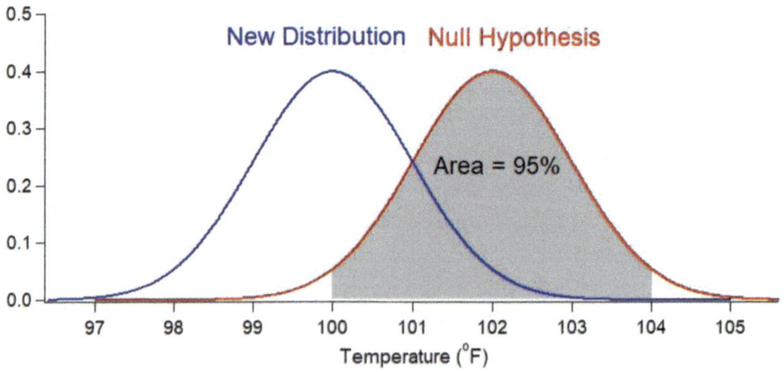

Fig. 3.1 Two normal distributions used to determine the significance of the drug Feverbreak. Both of the distributions are normalized Gaussian functions. The total area underneath each Gaussian is 100% and the shaded area under the original distribution, labeled 95%, has a width of two standard deviations. The standard deviation is $\sigma = \pm 1^oF$ so that two standard deviations corresponds to $2\sigma = \pm 2^oF$. The new distribution observed after giving the drug Feverbreak has a mean of 100 ^oF, which is shifted by two standard deviations from the mean of the original distribution at 102^oF

Table 3.1 Mock data for the test of a drug called Feverbreak on a patient population of twenty patients

Group/Individual	Fever	Treatment 2 h
1	102.7	101.3
2	103.0	100.6
3	103.4	100.4
4	102.9	99.0
5	103.3	98.6
6	101.2	98.6
7	101.6	99.8
8	103.1	101.1
9	101.7	99.4
10	103.0	101.1
11	100.5	100.9
12	100.3	99.5
13	101.6	99.9
14	101.1	99.4
15	102.6	99.5
16	101.3	101.2
17	101.5	99.8
18	102.0	98.9
19	100.2	100.4
20	102.5	99.9
Mean	102.0	100.0

(third column in Table 3.1). Figure 3.1 shows the comparison of the two distributions. In frequentist statistics, the new distribution is considered significantly different if its mean is more than two standard deviations away from the original mean. The area under a region that is bounded by two standard deviations on either side of the mean corresponds closely to 95% of the area under the distribution curve. We could also say that the probability of being more than 2σ below the original mean by random chance is about 2.5% (in this case more than $2^{o}F$ lower). The other 2.5% probability is for the possibility that the distribution is more than 2σ higher than the original one. That would also be significant, but significantly worse since it would mean a fever of 104 ^{o}F. In general, if we determine a mean value that is outside the 95% region, then we define that the difference from the mean value as significant. There is less than a 5% probability of measuring a value outside the shaded region by random chance. This is expressed as the p-test, which is also known as a significance test. The probability for the mean of the new distribution to be observed at more than 2σ from the original mean is less than 5%, which we write as a probability $p < 0.05$.

In the frequentist approach, we may reject the null hypothesis if the new hypothesis is significantly different (in this case significantly lower) than the original one. Can we conclude that the drug had the effect of breaking the fever? We must be careful. Rejecting the null hypothesis does not prove that Feverbreak is an effective drug. It only proves that it reduced the temperature of this sample under the conditions of the study. For example, we note that the new mean of 100 ^{o}F is still greater than the normal temperature of 98.6 ^oF. We might wonder whether this indicates a lack of efficacy. To infer that the drug actually breaks the fever and that this conclusion would be valid for an entire population should require repeat testing, control experiments etc. Frequentist statistics has often been applied in a sloppy way as an all-or-nothing approach, in which rejecting the null hypothesis is taken mean that the alternative hypothesis is proven.

Bayesian statistics starts with an estimate for the probability of the hypothesis. This is known as the prior probability. The prior probability is altered using the conditional probability [131] obtained from an experimental test, which provides the *likelihood* of observing the data using sampling methods. The product of the prior probability and the likelihood is then used to obtain a posterior probability. Here, posterior means "after the fact" or after the new experiments. A simple form of Bayes method can be derived simply by comparing the ratios of the two probabilities as follows

$$\frac{P_{posterior\,2}}{P_{posterior\,1}} = \frac{P_{likelihood\,2}}{P_{likelihood\,1}} \frac{P_{prior\,2}}{P_{prior\,1}}$$

where $P_{posterior}$ denotes the probability after making a measurement or performing an experiment, $P_{likelihood}$ the probability that a measurement will give a particular result and P_{prior} represents the initial estimate of probability that a theory is correct. If the probability of the new mean (i.e. in this case the lowering of the fever) increases we say that hypothesis has greater credibility. Confidence intervals are defined in frequentist statistics to define the probability of measuring a value within certain limits based on the distribution determined by experiment. Bayesian statistics tends to refer to credibility as the relevant yardstick. Bayesian statistics is relative and does not reduce the conclusion to an all-or-nothing statement. Instead, if the probability for one hypothesized value relative to another tends to increase as measurements are sampled, then the outcome has an increase credibility [132].

To implement Bayesian statistics we begin with a neutral likelihood ratio. We assume that the drug has a 50% chance of working and 50% chance of failing. The prior probability, the ratio of chance of working to change of failing, has odds of unity. This example will show that if the evidence strongly favors one credible outcome our initial assessment will not have a detectable effect on that outcome after testing. We now determine the conditional probability (likelihood) with and without the drug. We can take the data in Table 3.1 from 2 h after the drug is administered and infer a temperature of less than 100 °F means that the drug is working and greater than 100 °F means that the drug is failing to work. Using Bayes approach we can test both condition 2 (with Feverbreak) and condition 1 (without Feverbreak) by counting how many patients have a temperature of less than 100 °F for each distribution. We see that eleven of the values in the Feverbreak column are less than 100 °F (condition 2), while only three of the values in the original (untreated) distribution are less than 100 °F (condition 1). Using these numbers and Bayes formula [131] for the conditional probability we find $P_{likelihood\,2} = 0.16$ and $P_{likelihood\,1} = 9.5 \times 10^{-7}$.

Thus, starting with an assumed probability of 50% for each possibility, $P_{prior\ 1} = P_{prior\ 2} = 0.5$ we find

$$\frac{P_{posterior\,2}}{P_{posterior\,1}} = \frac{0.16}{9.5 \times 10^{-7}} \frac{0.5}{0.5}$$

Since $P_{posterior\ 1} + P_{posterior\ 2} = 1$ we find that $P_{posterior\ 2} > 0.999$ and $P_{posterior\ 1} < 0.001$. Note that changing our estimate of the prior probability to a significantly less optimistic assessment, e.g. $P_{prior\ 1} = 0.9$ and $P_{prior\ 2} = 0.1$, would have had little effect on the outcome in this particular example. The probability that the drug has an effect is high. In this case we do not say that the drug's efficacy is proven, but rather that the probability that the hypothesis that the drug breaks a fever is credible has increased because of the measurements.

The sample size is important in frequentist statistics since the assumption of a normal distribution is only valid in the limit of a relatively large number of data points. For example, Fig. 3.2 shows that the random numbers that make up the twenty temperatures in the fever column of Table 3.1 do not match a normal distribution very well. As the number of data points or individual measurements increases for a random sample, the distribution does increasingly look like a normal distribution. It is not always possible to make measurements on such large data sets.

Most often, in medical sciences one can only obtain a small number of patients for a study. To apply significance testing, we must *assume* that the data are random. The central limit theorem states that data will have a normal distribution for a sufficiently large sample. The sample distribution is always considered to be normal, but the parent distribution is a measure of the scatter in the data that may deviate from a normal distribution. Small sample size increases the error because of a lack of certainty in the parent distribution. Larger scatter in the parent distribution, which arises from a small sample size, leads to greater uncertainty in the true variance. In

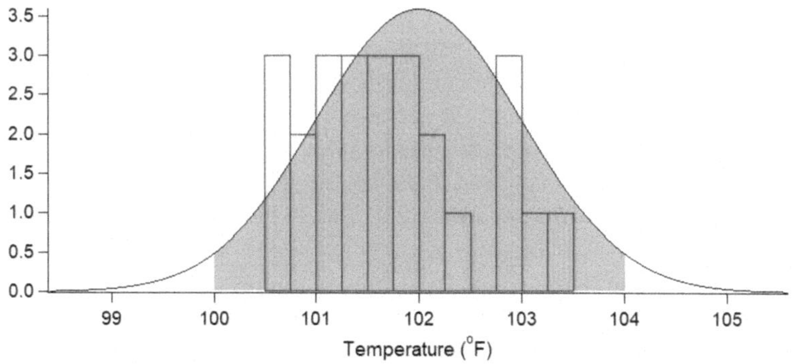

Fig. 3.2 Comparison of a histogram of the values for temperature in Table 3.1 compared to a normal distribution. Note that while all values fall within two standard deviations (the so-called 95% confidence interval), the shape of the distribution in not in accord with the normal distribution

1908 Gosset proposed a method to determine how much the width of the 95% confidence limit should increase for small sample sizes [133] Gosset published under the pseudonym Student as required by his employer, the Guinness Brewing Company. Gosset started with the theoretical width of a 95% confidence limit is $\pm 1.96\sigma$, which is expected if we have a large amount of data in a normal distribution. This is the theoretical value and it is close enough to $\pm 2\sigma$ that we used the $\pm 2\sigma$ value in the discussion above. Gosset made a table of effective widths of 95% confidence limit beginning with values as large as $\pm 2.73\sigma$ for a population of only three specimens. The appropriate width for the distribution decreases to $\pm 1.96\sigma$ in the limit of large numbers of data points (e.g. 100 or more).

A study on fruit flies may involve thousands of specimens and be completed in a week, but when we consider clinical trials on humans each individual may add complexity because of the specifics of their health history, the number of individuals may be quite small and the amount of time required to track the outcomes may be months or years. The mean does not have the same reliability in these two cases. Gosset's crucial test, which was popularized by Ronald Fisher and applied widely to Genetics and Biology [118], has come to be known as null-hypothesis significance testing (NHST). What started as a method for studying the raw materials used to make Guinness stout has to be used for making inferences on every subject from psychology to cancer biology [134]. Gosset warned about over-interpretation of the statistical significance tests that his method helped to make popular. Despite these warnings, NHST has come to be used by some scientists as an all-or-nothing evaluation of an experiment. For those who push NHST to the limit, if the statistical analysis shows significance (usually interpreted as $p < 0.05$) then the null hypothesis is rejected, and the alternative is taken as proven fact. This was never the intent of those who first proposed these methods.

3.4 Confirmation Bias: Seeing What You Want to See

Confirmation bias is a major problem in science relating to how hypotheses are stated and how control experiments are designed. Bias plays a large role in fields such as human health, psychology or social policy [135]. This is in part because these areas affect human living conditions more directly than chemistry or physics, but also because there are many variables that are difficult to account for. System complexity can provide an opportunity for bias, which is often unintended and obscured in how statistical methods are applied. Confirmation bias is also more prevalent when sample sizes are small, such as in studies of biology, living beings, and specifically humans. By contrast, measurements of atoms, molecules or physical phenomena arising from collections of small particles, from astrophysics to meteorology provide greater opportunities for control experiments, testing of fundamental physical laws, and robust statistics arising from large data sets. However, despite these differences, confirmation bias can arise in any discipline. It is "seeing what you want to see." Bias is often present in the experimental design and the assumption of randomness is often not appropriate [136]. It is easy to jump to the wrong conclusion because there are simply not enough data to substantiate a hypothesis. But a scientist working in the field will usually be looking for a result that can be published or used in a grant proposal. An even more dangerous practice arises when one searches for significance in a data set after an initial hypothesis has failed to achieve it (discussed later in Sect. 6.3 as p-hacking) [137]. Even if other biases have been removed or controlled, the need to publish a positive result injects bias into the study. To substantiate this point, a study of the p values of the publications in three different journals reveals a tendency of p values to cluster just below 0.05, the arbitrary cutoff for significance [138]. This observation suggests a bias that affects the inputs to the p-value at some point in a study. According to significance testing, a p-value greater than 0.05 means that the *null hypothesis* is true. This is often bad news for the academic looking for an effect of a drug, a treatment, or an environmental remediation to work. As long as the p-value is less than 0.05 the community will hail the result of the study as *proof* of the efficacy of a new drug, treatment or environmental clean-up strategy. So much is at stake in the study since often funding and promotion depend on getting a result. In this case 0.051 means the end of the research and 0.049 can mean champagne and a major award at the next conference. The observation that significance testing is failing to eliminate bias or appropriately control for small sample size has led some journals to ban significance testing [135]. In fact, some scientists feel strongly that this has become an ethical issue since scientists should know that the potential for self-deception or bias exists and is often built into a significance test [139, 140]. Even those who defend p-values do not deny that there is widespread abuse of these methods [141, 142]. There are several established methods that have been proposed to meet the criteria of a "new statistics" [140, 143] that would resolve the ethical and structural issues in significance testing.

Bayesian statistics can reduce confirmation bias by eliminating the all-or-nothing conclusion of frequentist statistics [132, 140, 144–147]. Using Bayesian methods one can gain confidence in hypotheses, using a quantitative approach that does not require either rejection or proof as the conclusion of a study thereby reducing pressure on the scientist to obtain a particular result [148]. Bayesian statistics is consistent with modern ideas in the philosophy of science.

3.5 The Descent of the Philosophy of Science from Rationalism to Relativism

Most scientists conduct experiments, read the literature in their field, and come to conclusions based on intuition and experience without concerning themselves about the epistemology of science. Many scientists take pride in their intuition and are weary of statistical methods. Gaussian or frequentist statistics, which we have called significance testing, is increasingly called into question. Bayesian statistics has gained ground as a foundation for the philosophy of science because it bears a similarity to the intuitive process that many scientists apply without formal training. In a manner similar to Bayes formal approach to conditional probability scientists intuitively assign each hypothesis a probability at each stage in the process of experimentation. There is seldom a full acceptance or rejection of a theory, unless that theory can be shown to be entirely baseless or falsified by some indisputable experiment or observation. Popper's example of black swans gives an example of a clear case for rejection. The claim that "all swans are white and have always been white" is proven false by the true statement that "In 1697, during the Dutch explorer Willem de Vlamingh's expedition, there were black swans on the shore of the Swan River in Australia." [149] Clearly, this is a limiting case. Rejection is usually not so clear cut. In fact, the difficulty in proving a hypothesis wrong is the weakness in Popper's proposal for how science progresses. For this reason, Bayesian statistics has become part of the philosophical view that has replaced the all-or-nothing approach, on which Popper's philosophy foundered.

The philosophy of science has treated the transition from one theory to another using various approaches. Popper reasoned that hypotheses could only be disproven, but not proven [150]. Kuhn focused on the subjective factors that resulted in theories being overturned [119]. Kuhn identified major changes as revolutions in human thinking starting with the Ptolemaic (geocentric) view of the solar system, then the Copernican (heliocentric) view, then Newton's mechanics, which explained the forces in the heliocentric solar system and finally quantum theory, which explained forces in the limit of small particles such as electrons, protons and neutrons. The subjective view of science has a very different meaning in post-modernist philosophy beginning with Feyerabend [151]. Post-modernism espouses the notion that the subjective experience of each person is equally valid and that cultural viewpoints of science should be given as much weight as the carefully developed body of research these comprise a field of science.

While scientists may find it difficult to take post-modernism seriously, they ignore these developments in intellectual circles at their peril. There has been alignment of trends within the society and the post-modernist faction in academia that has led to a backlash against certain new technologies and ultimately against science in general [152]. The idea that there is no objective reality became a popular view in some (but far from all) circles in social sciences starting in the 1970s. This philosophical viewpoint can be seen as a reaction to the influence of technological change on society and the environment. New technology can bring out a mistrust in a particular segment of the population [91]. Something as common today as the microwave oven was the subject a great deal suspicion and fear when it first became a household item in the 1970s. With time, microwave technology was accepted by most people and the subject of concern moved on to more recent technological advances. Today, there is suspicion of technology in certain segments of the population from cell phones and genetically modified organisms to electrical transmission lines. The suspicion may not be rational, but those who hold views that oppose scientific investigation often have a following and are effective at steering public opinion.

As a consequence of the breakdown in communication across disciplines, some of the post-modernists moved away from a rigorous foundation. To shine a spotlight on the lack of meaning in post-modernist ideas, Dr. Alan Sokal of New York University demonstrated that it was possible to publish an article in a respected journal that completely lacked rigor and scholarship. The article entitled "Transgressing the Boundaries: Towards a Transformative Hermeneutics of Quantum Gravity" was composed of intentionally written nonsense in a pseudo-scientific style. The article was published in the journal, Social Text, in 1996 in an issue devoted to understanding relativism in science [153]. The article was exposed by the author as a hoax shortly after it was published [154]. The point is that tolerance for subjective viewpoints to permit cultural diversity does not mean that we should uncritically accept all viewpoints on medicine, environmental science, chemistry and physics as having equal validity. However, scientists should not be smug about an incident that exposed the fallacy of post-modernism. Despite the rigors of their training, scientists have succumbed to numerous hoaxes and even outright fraud. Various disciplines can only learn from each other if each is willing to try to understand the foundations of the other. We must do the hard work of learning the how the intersection of knowledge in various fields may teach us how to confront societal, environmental and medical problems (to name a few).

Epistemology should be part of scientists' training in ethics. The fact that the funding agencies and modern university administration is focused on applied science tends to remove scientists from the historical antecedents and social context of science. The increasing difficulty of obtaining funding for basic science, as opposed to applied science, is having an effect on every aspect of our educational system. Because of specialization and simultaneous lowering of standards, student training has become increasingly vocational in the hard sciences. This is unfortunate because respect for intellectual traditions has been lost both by those who see scientists as an elite and by the self-imposed isolation of scientists from other forms of knowledge. Scientists' general lack of training in other disciplines is nowhere more troubling than in the ethical arena.

3.6 The History of Scientific Progress and Acceptance of New Ideas

We need to consider some examples of the role played by integrity in scientific breakthroughs to put scientific aberrations in the proper context. As the Copernican revolution was taking shape, the details of the planetary orbits in the heliocentric model were being worked out by Johannes Kepler using Tyco Brahe's data [155]. Although Brahe had the best data and provided it freely to Kepler, ironically, he did not believe the heliocentric model. Here is an example where two scientists had opposing views and yet they collaborated to advance understanding. While Kepler accepted the heliocentric view and he also believed that the orbits of the planets around the sun would be perfect circles, as part of his belief in the "harmony of the spheres". One of the reasons we see Kepler as such a great scientist is that he faithfully followed Brahe's data and came to the conclusion that the orbits of the planets are, in fact, not circular, but rather are elliptical [156]. Competitors at the time published tables of the motions using a circular model, which were easier to understand, but not accurate. It would have been so much easier for Kepler to ignore the small differences and to follow the trends of the day. But he followed the data instead. Today Kepler's elliptical orbits are the accepted theory and his steadfast approach in following the data, even when it did not conform to his bias, is exemplary and even heroic. We must remember that the Catholic church did not readily accept any of these ideas and scientists faced a great deal of strife in their personal lives in addition confronting ethical issues in the detailed treatment of data [155].

The following topics that relate to the technological claims fundamental to the hexagon case. Each of the following examples makes a point about how a particular scientific discovery was received by the community. There are notable examples of completely unexpected breakthroughs that were readily accepted by the scientific community because the evidence was considered overwhelming. Acceptance is easiest when the observation comes out of the blue and does need to replace an existing theory. High temperature superconductors are a good example of an unexpected discovery. In 1911 Kammerlingh-Onnes discovered superconductivity, which is the conduction of electrons with no electrical resistance. This was in itself a surprising finding, but for 75 years it was considered a curiosity observed only at very low temperatures [157]. Under normal conditions at room temperature, the motion of the nuclei of conductors (e.g. metals) impedes the flow of electrons, which gives rise to electrical resistance. One can understand this resistance because of the electrons bumping into the nuclei as they move along a conductor in response to an electric field. In a superconductor the electron and nuclear motions are coupled, which permits electrons to transit through the material essentially without bumping into the nuclei. For 75 years following its discovery, the only superconductors known were metals such as mercury or niobium that required temperatures for operation only a few degrees above absolute zero of the scientific temperature scale at −273 °C (−459 °F). Hence, practical applications of superconductivity were limited. The possibility of higher temperature superconductors was discounted by many scientists as theoretically impossible, until the effect was unexpectedly demonstrated by two researchers at IBM, Karl Müller and Johannes Bednorz in 1986 [158]. The high temperature superconductors still operate well below room temperature, but in a range that is much more readily accessed. The high temperature superconductivity project was carried out by these researchers in industry, rather than a university, in part because the skepticism of the scientific community made funding such a project all but impossible in an academic setting. When the first high temperature superconductor was discovered it completely changed the way scientists viewed the entire class of metal oxide materials and the possible technological applications of superconductivity. Recently, room temperature superconductivity has been observed at very high pressure. Such discoveries excite the imagination. It appears increasingly likely that a viable superconducting technology could become a reality.

When the two professors who reported the hexagons marketed their grant proposals, they proposed that RNA could be used to discover room temperature superconductors. Since high-temperature superconductors were completely unexpected, their salesmanship appealed to the hope that a next breakthrough could lead to room temperature superconductors. This hope was one of the crucially important proposed ideas used to justify the concept of RNA-evolutionary materials synthesis. Incidentally, the recent discovery of room temperature superconductors at extremely high pressure would never have been achieved by the path proposed by the two would-be inventors. The hexagons were presented as the first example of something much larger, an evolutionary method to develop new materials, but there was never any evidence that the hexagons were superconducting, catalytic or indeed had any interesting properties of any kind. Their grant proposals on so-called RNA evolutionary materials synthesis were pure marketing with no data other than images of a few shapes under an electron microscope. The hexagons by themselves would most probably not have been a high-value discovery, even if they had been composed of pure palladium metal. It was the promise of a breakthrough of the same type as high-temperature superconductivity that was crucial to the marketing in the proposals. Their marketing relied on the convincing reviewers that a combination of an evolutionary approach based on a selection procedure using RNA as the means to achieve the desired breakthrough.

To understand why RNA was part of the marketing we need to return to the 1980s when scientists began to manipulate genes by use of DNA made in the laboratory. During the same period, a second breakthrough in cell biology came when scientists began to understand the regulatory role of RNA. DNA is a biological and chemical information storage repository in a living cell. Before 1980 RNA was known as messenger RNA that carried copies of DNA to a cellular factory called the ribosome, which is also made of RNA called ribosomal RNA. The ribosome is a cellular "factory", which matches transfer RNA, each of which carries a three-letter code, to a complementary section of messenger RNA (see Fig. 3.3). The matching of the code on the transfer RNA to that on the messenger RNA shown in Fig. 3.3 guides a specific unit of a protein into place in the process of protein synthesis.

In 1989 Drs. Thomas Cech and Sidney Altman received the Nobel Prize for showing that RNA can also act as a catalyst outside the ribosome [159, 160]. The ribosome is the factory that translates RNA sequences into protein sequences in living cells. Their research showed that certain RNA sequences can chemically cut and splice other RNA, like editing a text. This means that RNA can function in the processing and regulation of other RNA to alter the sequence of Messenger RNA [62, 160, 161]. The old hypothesis of [DNA ➔ Messenger RNA ➔ protein] is the so-called central dogma of biology. The new view can be summarized as [DNA ➔ RNA(transcript) ➔ RNA(editing) ➔ Messenger RNA ➔ protein]. This breakthrough led to an understanding of a whole world of RNA interactions that has implications for medicine, agriculture and for our understanding of life itself. In the old view, one DNA codes for a specific protein. In the new view, one DNA gene can code for many proteins and the regulation is at the level of the RNA. DNA technology has become essential for making new proteins and for the possibility of

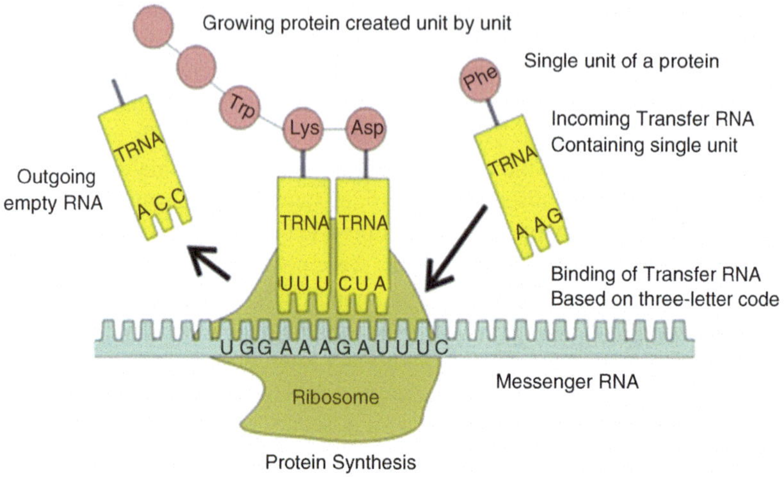

Fig. 3.3 A schematic representation of the process of translation, by which the sequence of messenger RNA (derived from DNA) is translated into a growing protein. Protein synthesis occurs by the addition of specific units coded by the three-letter code of transfer RNA that matches with the complementary code on messenger RNA. There are twenty different types of units and a code of 64 possible three-letter words. There is redundancy in the code

introducing genes to treat diseases (gene therapy). However, new catalytic RNAs may have even greater potential for therapy and may help switch production of one type of protein for another by "editing" the Messenger RNA and changing the series of "letters" read by the ribosome and translated into protein. In this case, there was some resistance to the new ideas, but it lasted only a few years as Cech responded to criticisms and developed experiments to prove the concept of RNA catalysis. In this description, we would not say that the old "central dogma" was an error, but rather that it was a simplified view of nature that was refined by the discovery that RNA had many functions that had not been previously known. This example also has relevance to the hexagon case since the marketing of the idea that RNA could make new materials was one of a series of proposals of how to use new functions of RNA following Cech's discovery. Many of the ideas developed into new technologies, but that does not mean that all ideas are founded on solid ground.

It is important to present an example of how the scientific community deals with errors in scientific publications in a proper manner to provide some background for many of the other examples discussed. In the quantitative sciences of chemistry and physics, examples abound of specific instances where long-held truths were upset by studies conducted with better instrumentation or more complete control experiments. The structure of carbon monoxide bound to the iron atom in human hemoglobin as a specific example of a long-held model that was completely changed because of new findings. Hemoglobin is the most abundant protein in red blood cells, whose function it is to carry oxygen from the lungs to other tissues in the body. The hemoglobin protein contains a relatively large flat molecule called heme, which is responsible for the red color of the protein. Hemoglobin was the first protein whose structure was solved by x-ray crystallography. The structure provides a three-dimensional picture of hemoglobin, both with and without bound oxygen. In 1962, Max Perutz and John Kendrew jointly won the Nobel Prize for the determination of the structures of hemoglobin from blood and myoglobin from muscle [162]. Hemoglobin and myoglobin are closely related in structure, readily available from blood and tissues, and are readily crystallized. The formation of protein crystals, and hence the possibility of the determination of their structures by x-ray diffraction, was pursued for many proteins in subsequent years. There are tens of thousands of protein structures available on the publicly supported Protein Data Bank (http://www.rcsb.org/pdb/home/home.do). Using the PDB website one can see a picture of the structure of the proteins that have been characterized. In most cases, x-ray crystallography is one of the most reliable techniques for structural information in scientific investigations. The evidence of x-ray crystallography is often considered as more informative than other measurements because it provides a three-dimensional picture of the geometry [163]. However, despite the common perception among scientists that x-ray crystal structures are more informative than other methods, there are numerous examples where other methods turned out to at least as important as x-ray crystallography in understanding the bonding of molecules in a protein. The structure of carbon monoxide bound to myoglobin is one of those cases.

Based on the x-ray crystal structures of myoglobin and hemoglobin determined in the 1960s, it was known that the oxygen we breathe binds to the heme iron in a bent geometry. This bent geometry is observed in proteins, such as myglobin or hemoglobin and in structures of isolated hemes as well. On the other hand, free heme structures with bound carbon monoxide show a bond angle of 180°, the molecule is not bent, but rather it is perpendicular. Despite this fact, for more than 30 years, the structures of myoglobin and hemoglobin that contained carbon monoxide showed a bent geometry. This would not be expected on chemical grounds and is clearly at variance with the structure of a free complex (without the protein surrounding the heme) [164]. However, the bent heme carbon monoxide structure in myoglobin fit nicely with a bias that many scientists had. Scientists came to believe that the globins enforced the bent geometry for carbon monoxide in the same way that oxygen was bent. One prominent theory was that the forced bending discriminated against the binding of carbon monoxide and in favor of oxygen. The discrimination in favor of oxygen binding is a good thing since there is some naturally occurring carbon monoxide in our bodies and it would be lethal if carbon monoxide were to bind to hemoglobin and myoglobin with the same affinity that it binds to free heme. For decades, chemists, physicists and biologists were satisfied that the bent carbon monoxide myoglobin structures showed a "strain" induced by the protein that caused the carbon monoxide binding to be weakened. The picture was so much a part of the mainstream that it was presented in biochemistry textbooks. But the structure showing a bent Fe-C-O was completely wrong. In 1994 the textbook picture of the binding of carbon monoxide and oxygen to hemoglobin had to be changed based on an experiment using polarized infrared light that showed that carbon monoxide is perpendicular, and not bent [165]. Very soon thereafter, these results were confirmed by structural studies of the protein using the best methods available at synchrotrons, which were not available in the earlier studies [166]. By 2000 the definitive conclusion was that the carbon monoxide in myoglobin was perpendicular and not bent. These results disagreed with 30 years of scientific thinking, and yet in the face of compelling data, the implications of the new structure were immediately accepted by the scientific community [167]. This highly specific example serves to show that an incorrect hypothesis can survive for many years even with relatively good data. More important is the fact that scientists continue to search for the truth even when an answer has been previously found and is widely accepted. It is a hallmark of proper science that its practitioners are skeptical and continually reinvestigate even well-established results. It is important to have such an example to show that scientists can discard entrenched ideas when conclusive evidence has been presented to establish that science can and does function this way a great deal of the time. But science does not always function in this way.

3.7 Black Boxes: The Abuse of Technology

As science has advanced, automated instrumentation and computers have become a valuable tool to advance knowledge. However, the increasing sophistication of the available tools has been accompanied by the abuse of instrumentation and computer programs to obtain meaningless results. When discussing the use of computer programs for calculation of electronic structure, solving x-ray crystal structures, interpreting high resolution scattering data or other advanced methods, students are often warned about treating the instrument as a "black box". The term refers to use of an instrument without really understanding its capabilities and proper use, settings, calibration, or proper interpretation of the output. With such powerful methods at our disposal it is surprisingly easy to produce results that have no significance. In the world of computers, the common adage is "garbage in, garbage out". Your scientific output is only as good as the thought given to setting up the problem and implementing the calculation or measurement with all proper settings. Scientific training is supposed to prevent abuse of this type, but there are lapses. It is so difficult to understand all techniques that are available that errors are bound to occur. Small self-reliant research groups can sometimes foster a culture of careless use of powerful tools that are available in university facilities. How can we reign in the abuse when it occurs? Examples of attempts to correct the errors reveal a stubborn resistance on the part of those who have followed poor practices. No one likes to be told that they are mistaken.

There are numerous examples in the field of protein x-ray crystallography, in which there is a significant subjectivity in data interpretation. Many protein crystal structures are studies of drug binding or "ligand" binding to proteins. The drug or ligand is a small molecule, but still quite a bit larger than carbon monoxide discussed above. In the interpretation of the data one is often looking for the molecule in a rather subjective way by attempting to fit the molecule into a blob of "electron density" at an intermediate stage in the analysis. Determination of an x-ray crystal structure requires that the diffraction data be "indexed". The indexing process identifies a unique set of parallel planes running through the crystal that give rise to each diffracted spot. The diffracted spots are transformed into a model of electron density by a mathematical process. Using a graphical computer program to visualize the shapes, the protein is positioned in the electron density to attempt to account for the electron density in an iterative process, using the known protein sequence as a constraint. Although the protein is constrained by its linear sequence and certain three-dimensional structures, a small molecule such as a drug molecule does not have the same constraints. Thus, there is a subjectivity in the positioning of the drug molecule in the density to build a model. This subjectivity is precisely why it was possible to assume that carbon monoxide had a bent geometry in the muscle-storage protein, myoglobin for 30 years [166, 167].

Added to the complexity of model building is the consideration that a drug molecule may not be bound to 100% of the proteins in a crystal. As a result of the experimental uncertainty in the occupancy and position of a drug molecule it is possible to position a drug molecule in a small blob of density and justify the model by saying that only a small percentage of the proteins have drug molecules bound to them. Such a procedure can result in a completely erroneous model. The need to introduce standards into the protein x-ray crystallography was noted by Kleywegt and Jones [168–170]. In a paper entitled "Where freedom is given, liberties are taken" these authors document errors in interpretation of electron density in automated structure determination [168]. This issue was considered systematically by Dr. Bernard Rupp and co-workers in the assessment of small molecule, drug or ligand structures directly from the deposited x-ray data [171, 172]. There are many x-ray crystal structures of protein-drug complexes that are incorrect in the sense that there is no real evidence that a drug molecule is present in the structure [173–176]. Given that there are financial incentives in many protein-drug complex structures some research scientists object to the use of their data to reanalyze the results even though they have published and placed their data on a public database [173, 174]. The climate has changed rapidly in science and the collegiality that used to be commonplace is being replaced with a culture of privacy and legal protections. The highly competitive funding climate has a negative impact on open debate, and on ethics.

3.8 The Sociology of Science: From "Follow the Data" to "Follow the Money"

The notion of science as a vocation, a higher calling, that should last a lifetime was enunciated by Max Weber [177]. This idea has become part of the popular myth about who scientists are. It is a flattering vision of a group of people dedicated to solving problems that plague humankind and solving mysteries that have perplexed humans throughout history. Sociologist Robert Merton set out to study the sociology of science as it emerged from World War II with a mandate to expand its role in society. Merton proposed four norms that describe the social fabric of the scientific community; communalism, universalism, disinterestedness and skepticism [178]. Financial considerations are completely absent from the Mertonian norms or his work on the scientific reward structure. Thus, Merton's social theory ignores one of the most crucial aspects of modern science, which is that the cadre of scientists are career professionals whose livelihood depends on the success of their scientific work in garnering grant support. Merton saw scientists engaged in a process of working towards a scientific discovery that was inextricably linked with ego and the desire for recognition. Merton's norms are dated and should be reconsidered in the context of the increasing trend towards privatization of the university system.

Merton's ethos of science has been used to support the idea that science is self-regulating. Many scientists have internalized this ethos through their informal training and discussions in a research group about what is means to do science and to be a scientist. The narrative of self-regulation implies that scientists will recognize mistakes because they will lead to conclusions that are contradicted by other experiments that have been done correctly. Those who made the mistake will then go back and determine their error, fix it, and report the correct procedure. Indeed, many scientists do practice science with such rigor, but that does not mean that science *always* works in this way. There are self-deceptions, in which scientists refuse to see a mistake because it is inconvenient, perhaps because it has already been published, or because it would invalidate a funding proposal, or some other reason that has to do with the survival of the research group. Finally, there are falsifications or fabrications that result from intentional fraud.

A scientist may face an ethical dilemma if she or he finds evidence that contradicts her or his own hypothesis. As contradicting experiments are published one approaches the difficult question: at what point should a scientist abandon or modify a theory? Those who are holding on to a theory despite mounting indications to the contrary, may say that the evidence is not yet complete or that there is still some likelihood that further experimentation may change the interpretation to favor the theory once again. But scientists can struggle even in cases were matters are clear. Why? Because scientists may be influenced by social, psychological, and personal factors that have nothing to do with the pursuit of truth. For example, a scientist might require more countervailing evidence than he otherwise would before rejecting a theory if his or her career depends on that theory. If abandoning the theory would precipitate a crisis in funding crucial to the research group, it may be difficult for a scientist to see the likelihood that evidence implies the failure of theory. An outsider whose funding does not depend on the result may have an easier time accepting the evidence.

Traditional academic culture has been upended by changes in university funding. Merton's school of thought characterized by a "follow the data" approach has been replaced by a pragmatic "follow the money" approach. The commercialization of science and the mixture of corporate and academic culture are two aspects of the pressure placed on scientists in the modern university. A large shift in priorities occurred when funding for nanotechnology was announced. The National Nanotechnology Initiative of 2000 was legislation that required funding for nanotechnology to increase by 1% per year in the budget of the NSF, NIH and other science funding agencies [179]. The shift in priorities of scientific research followed suit and nanotechnology appeared everywhere. These was an explosion of nanotechnology journals; nanotechnology centers; and old technologies were renamed nanotechnology to try to position scientists to be competitive for the new funding source. The trend towards nanotechnology funding has been intertwined with a concerted effort by universities to build technology incubators and jump-start the formation of companies. These initiatives have not been led by scientists and are not necessarily beneficial for scientific priorities. The shift in funding caused a dislocation and forced some scientists to rethink their research programs. The requirement to show a nanotechnology connection in a particular research effort increased the space devoted to marketing increased at the expense of rigor of scientific thought.

3.9 The Marketing of Science and Scientific Branding

The increase in corporate funding and patenting in universities has permeated grant writing, with an increasing emphasis on convincing peer reviewers, investors and the public that science is a utilitarian undertaking. Given the importance of assigning credit to scientists for their contributions it is crucial to have an appropriate system for evaluation of the products that define a scientist's output, publications, and grant proposals. Peer review is considered the best method for review of either publications or proposals. Often the reviewers of manuscripts are referred to as referees whereas the grant reviewers are referred to as reviewers or a panel, if one refers to the group of reviewers as a whole. Authors send manuscripts intended for publication to a journal editor who will then select referees. The referees are always anonymous, but most journals will provide the names and affiliations of the authors, so that most peer-review is a single-blind system. Double blind manuscript review is rare, but many scientists have begun to raise the issue of the need for a less biased system. Double blind review would be a possible innovation to provide greater fairness in reviewing. Grant proposals sent to the various funding agencies can be sent to individual reviewers or assigned to a panel for consideration. The program manager always has some input. In some agencies, the program manager makes the funding decision, and the review process is purely advisory. Even in the NSF, where peer review is stated to be the major criterion, non-scientific criteria may also be used in making the final decision. Finally, some funding agencies leave the funding decision up to the discretion of the program manager. Assembling peer reviewers and bringing them together for panel review is a grueling experience in which groups from 4 to 30 participants discuss the various proposals and rank them on their merits. Then those proposals that have scores above the "pay line" are funded. Studies of peer review have shown that the ranking of proposals varies greatly with the composition of the review panel. In one of the few studies of this process, the outcome was more random than researchers had anticipated [180]. Scientists often comment that submitting a proposal feels like playing a roulette wheel. It is depressing for them that the research supports that hunch. On the other hand, there are stars in academia who are recognized and those who are part of the elite are much more likely to receive funding. Research suggests that elitism is personal and not institutional [180]. Regardless of reasons for the star phenomenon, funding tends to be concentrated in a few large groups rather than spread evenly across the scientific community. These factors result in a well-known tendency to disfavor young scientists. This fact is known to the funding agencies and has led to special funding categories for those who are starting out. These programs have had mixed success [2]. Since funding is the single most important determinant of a scientists career the situation is not very encouraging. A person can work extremely hard only to find that they are denied funding and the opportunity to continue. Despite the research described above, decisions are *supposed* to be made based on the content of proposals and the productivity of the scientist. The reality is that many other factors enter into funding decisions.

How do scientists assess productivity and quality of achievement in an increasingly specialized world? Since it is not feasible to read all publications and evaluate them in the context of the particular discipline, scientists have resorted to a more superficial assessment, by counting the publications and weighing them to some extent by the impact factor of the journal. Impact factor is a number that tells one how well cited the journal is. It is also important to establish that the papers have an impact, that someone actually reads them. Since there is no way to know whether scientists actually read the articles, the system that has developed is based on how many times a publication is cited by other publications. This leads to a citation index for each scientist and way to ascertain the impact factor of a journal. There is a pyramid in the citation index, with a few groups getting most of the citations, but does that mean that the rest of the publications are not worth reading? [181] "The Ortega hypothesis" suggests that a large group of scientists with low citation indices pave the way for few to make even greater progress, i.e. that their work really is necessary as a base of support for the more productive scientists. However, a research study failed to support the Ortega hypothesis. The research supported the benefit of concentration of funding in the hands of a few scientists [181]. On the other hand, the hype about publications and citation indices completely misses one of the most important aspects of research, which is the training of the next generation of scientists. Every professor understands that graduate students will choose between being in a large group with high prestige and relatively little interaction with their mentor or a small group with lower prestige and a great deal more personal mentoring. Both have advantages and students distribute in some measure in similar proportion as the publication output of groups (which is not a surprise since the students are responsible for that output). Unfortunately, the training dimension is hard to measure and is all but absent in the metrics used to evaluate research scientists.

In recent years, the metric of a scientist's output has been reduced to a single number, the h-index. The h-index (or Hirsch index), is calculated based on a rank order of the number of times an author's publication has been cited. One goes down the list until the number of citations is equal to the number of the publications. For example, a scientist who has twenty publications that have been cited at least twenty times has an h-index of 20. The reduction of the entire lifetime output of human being to one number is tragic development of the modern university. But this is where we are today in a world that is so specialized that scientists have given up trying to understand each other's work or its significance. They simply look up the h-index to see who is who.

Increasingly, scientists find themselves in a straitjacket of metrics rather than the world of ideas and insights. Metrics such as the h-index further encourage the current trend to base promotion and funding on "productivity" quantified in a way that lacks meaningful consideration of the intrinsic merit of the work. However, a scientist who makes a widely recognized breakthrough can transcend the tedium of academic ranking that pervades contemporary scientific culture. Whether it is for money or for fame, the hope for a breakthrough motivates many scientists. The pressure to distinguish one's science by making a "breakthrough" increases the possibility of mistakes. Scientists are prospectors constantly looking for the interesting insight or result that will appear when they least expect it. However, one must always be wary that the unexpected result could really be a mistake that may have arisen owing to the lack of a control experiment or a condition that is not properly defined. Scientists must always be wary. If something looks too good to be true, it probably is.

Chapter 4
The Corporate University

> *The ideals of scholarship are yielding ground, in an uncertain and varying degree, before the pressure of businesslike exigencies, a merchantable commodity to be produced on a piece-rate plan, rated, bought and sold by standard units.*
>
> *– Thorstein Veblen, The Higher Learning in America, 1918*

The hypothesis that universities have a profound conflict of interest when they are called upon to investigate their own faculty is based on the funding models that sustain them. The loss of federal funding alone would be a significant downside to finding that a faculty member had committed research misconduct. However, many of the most controversial and risky projects are also linked to private funding models. The increasing dependence on private funding at public universities has further exacerbated the contradiction embedded in the federal research misconduct regulation.

Certain aspects of the trend towards marketing science and development of the university as a corporation has antecedents that date back even before World War II. The role of science education in the United States was given a significant boost by the 1862 Morrill Act, also known as Land Grant legislation that set up agricultural extension as a major justification for expansion of the educational mission of colleges in many states. The law called for colleges with a focus on agricultural and engineering for men and general science and home economics for women. With time these colleges evolved into the land grant universities with an initial significant input from science that boosted production and profitability of the farm sector economy. From the beginning a focus on science education with a practical orientation was deemed essential in American universities. In the early years, science education focused on training with only limited resources devoted to research. However, commercial applications were almost always developed outside of the Ivory Tower. The university's core mission in the early 1900s was geared towards the education of an informed citizenry. The idea of patenting research was considered inappropriate or even unethical for the first 50 years of the Land Grant university system. But over the years attitudes began to change.

S. Franzen, *University Responsibility for the Adjudication of Research Misconduct*, https://doi.org/10.1007/978-3-030-68063-3_4

There were only a small number of patents issued to universities prior to World War II. The proceeds went to foundations and to the university rather than to individuals. But even so there were observers who questioned the legality of this procedure [182]. In Europe and the United Kingdom patenting medicine was strictly forbidden [183]. But the situation in the US was less structured and varied from university to university. The research done at universities with public funds was seen by many as incompatible private ownership and use. The change in status began with formation of the Wisconsin Alumni Research Foundation (WARF), founded in 1925, which is seen as the precursor to the modern Technology Transfer Office (TTO) [184]. The foundation was initiated to accommodate patents on the enrichment of vitamin D based on research done at the University of Wisconsin (UW) by an professor/inventor named Dr. Harry Steenbock. The 1923 insulin patent of the University of Toronto was another early exception [182]. On the other hand, Harvard University had a strict policy forbidding patents and many other universities followed this policy. The proceeds from the few patents that existed went to the universities and not to the inventors. Essentially, the funds became part of an endowment and with the justification that they benefitted research and infrastructure projects for the public good.

The contrast with today's university is striking. From the modern perspective we might call UW one of the "success stories", but at the time patenting anything from university research was so controversial that UW was more vilified than celebrated. The change in attitude occurred during World War II as a result of the development of penicillin. This massive project involved transfer of technology first from the U.K. to the U.S. and then involved collaboration of companies and universities to work towards mass production of the drug. The wartime imperative of developing an antibiotic is obvious and perhaps the needs of the war justified a change in research strategy, as well as a step in the direction of privatization of research. Companies and universities were enlisted by the government in an effort that involved more than 1000 scientists. The successful project created a paradigm that was later used to promote greater collaboration between the private sector and universities. However, the idea lay dormant for more than three decades after World War II. In 1944 the vitamin D patents were revoked by a court decision, in which the judge rebuked the Wisconsin Alumni Research Foundation (WARF) for facilitating university research in collaboration with pharmaceutical companies. This was a stunning decision when seen from today's perspective [185]. The idea of university-owned patents took a long time to mature.

The report entitled "Science – The Endless Frontier" written by Vannevar Bush for President Truman in 1945 has been seen as an antecedent to modern technology transfer. Indeed, Bush had been the director of the wartime research funding agency that had coordinated the development of penicillin, which used such a model [186]. However, Bush believed that the freedom to pursue fundamental science would lead to great benefit for society. Thus, the Manhattan-project style of science was set aside, and the National Science Foundation was created to fund small research groups at America's universities. One can view the post-war period in the United States from 1945 to 1970 as the golden age of basic research. As it became clear that

American science had become preeminent in the world, the funding for science from the federal government increased steadily. Universities have become reliant on the proceeds of that funding, which has provided more than just research support for scientists. There were important structural changes that followed the rise of funding for science. One crucial change is the role played by grant overhead, which provides a cash flow for the university administration. The second important change is the transformation of universities into engines of economic development [185].

4.1 The University Funding Model

Universities charge an internal tax on the funds received by scientists from external funding sources. This tax is known commonly as "overhead" or "indirect cost" in the official language of the administration. The proceeds from indirect cost are used for maintenance of buildings, services, libraries, computers, infrastructure, and support staff, but are also used to support the administration at large. This model for funding the university has become essential to the modern university in the U.S. Its importance has increased as state support of universities has declined. Indirect cost has an average rate of ~48% of the grant amounts based on data from the Council on Government Relations [187]. Rural universities have the lowest overhead rates (~28% in rural Kentucky) and urban universities have the highest rates (~69% in New York City). National laboratories have even higher overhead rates rising well above 100% in many cases. Thus, for every dollar of taxpayer money spent on research administration, services and maintenance can cost the same or more. The grants obtained by scientific research personnel, professor or staff scientists support the institutional cost of doing business. Scientists must justify their positions in academic, private or government research institutions by raising grant funding that pays for their research and often at least part of their own salary.

The flow chart in Fig. 4.1 describes the process by which science is funded and how overhead then flows to the coffers of the university. Each funding agency shown has a distinct mission. For example, the NSF supports fundamental scientific research and education in institutions of higher learning, but also has a societal mission to promote understanding of science and to support K-12 science education to help train the next generation of scientists. The National Institutes of Health (NIH) funds research to discover new treatments for disease as well as methods of prevention. NIH funding spans research areas from fundamental studies of methodology, e.g. individual proteins and DNA/RNA structures, on the one hand, to clinical trials and the testing of novel therapeutics at the other extreme. The Department of Energy (DoE) funds research into energy conversion technologies that include traditional energy sources such as petroleum products, coal and natural gas, but also alternatives that include solar energy, and wind energy and so on. The Department of Defense (DoD) funding focuses research into new methods to support the mission of soldiers in the battlefield and advanced technology for wound treatment, threat detection, secure communications, missile defense, and environmental restoration,

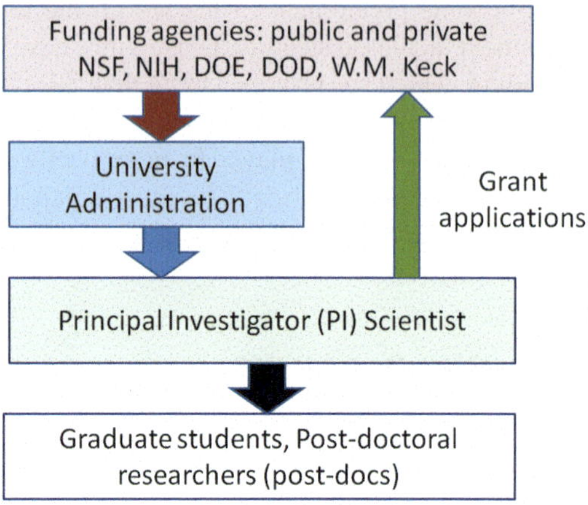

Fig. 4.1 Funding flow chart for academic research. The source of support for most research from is federal or private funding agencies, such as the National Science Foundation (NSF), National Institutes of Health (NIH), Department of Energy (DoE), Department of Defense (DoD), or private foundations such as the W.M. Keck Foundation shown in the top panel. The figure shows the flow of funding from agencies to universities and then principal investigators (PIs). Once a proposal has been accepted by an agency, funding is dispersed to the university. The university administration retains a portion of the funding known as indirect cost or overhead. In return the university provides infrastructure and other support services for the scientists. The PIs use the research funding to purchase equipment and supplies and to pay the stipends of graduate students

to name a few. There are some cross-cutting areas such as environmental research, which are part of the portfolio of many agencies. Each of the federal agencies has a different review process, but they all involve a program officer who is in charge of a funding portfolio in a specific area. Private funding sources also provide support for university research. For example, the W.M. Keck Foundation funds highly novel or ground-breaking research, but also supports buildings and major infrastructure at prestigious universities.

The ownership of intellectual property and a new relationship with industry has transformed the American university in recent decades. Historically, universities subsisted on a combination of tuition, donations, and public funds. This equation changed in 1980 when the Bayh-Dole act allowed universities to become the owners of intellectual property, which further permitted them to license patents and to own equity in corporations [186, 188]. This legislation has far-reaching implications for the university's role in society. Prior to Bayh-Dole academic research played a peripheral role in development efforts in established industries. Bayh-Dole ushered in a structural change whereby universities have a substantive stake in emerging industries via intellectual property rights and actual equity in spin-off companies. With that role comes a responsibility to ensure that there is a clear public benefit to the research output. Under Bayh-Dole, the university can manage an intellectual property portfolio just like a corporation. This new role has an insidious effect on the power and responsibility felt by a university administration. The new economic mission of the university has been used to justify the shift from basic research and education by research-active faculty to less expensive and more expedient contract research and non-tenure track teaching faculty.

The overall stagnation of public funding for research drives universities further down the path of commercialization in the hope that the new funding sources will pay off. When seen in aggregate, according to the Congressional Budget Office the total federal support for scientific research as a percentage of GDP crested in 1962 at ~1% and has declined to around 0.5% today [189]. Since overall research support has fluctuated between 2.3 and 2.6% of GDP for the past 30 years phirivate sources have made up the difference rising from 1.3 to 2.1% of GDP over the same period. Research support by industry and university-sponsored spin-off companies has increased in importance as the federal portion of the funding has decreased as percentage of GDP. The decrease this governmental investment in research runs the risk of having a negative long-term economic outcome. Much of the current prosperity in the United States has been propelled by the governmental investment in everything from penicillin and the polio vaccine on the biomedical side of research to the transistor and the internet in the tech sector. As companies have downsized their research and development efforts, they explored academic partnerships. However, industry-sponsored research has remained remarkably constant over a 30 year period following the passage of the Bayh-Dole act in 1980 and accounts for 5–15% of total research funding [2, 190]. Corporate sponsored research is basically contract research, which may actually be threatened by the universities' rights to own patents. Companies do not want to lose their patent position simply to gain a modest benefit from collaboration with a university research group. The patenting of academic research output increased significantly throughout the 1970s and was perhaps the driving force for the Bayh-Dole act [191]. Patenting results in the potential to provide a large payoff for the university in the form of licensing of patents, royalties from product sales and equity in companies. However, by the estimate of a recent Brookings Institute study, most universities are losing money on technology transfer [192].

The notion that spin-off companies and patents can replace other funding sources in the educational arena is reminiscent of the debate over academic sports as a source of financial support for universities. In each case, a large investment is required. Infrastructure is the major investment for college sports, although coach and staff salaries can also be significant. This is often justified in terms of revenue, but it is not clear that sports generate significant funding for the educational mission of the university. Rather sports generate interest and feelings of support that lead to donations, which can, in principle, be used to support education. In a similar way, management of intellectual property and spin-off companies requires a significant investment in administration to maintain a competitive edge for the university [193]. A considerable up-front investment is required maintain a TTO with highly-paid administrators, legal contracts for patenting and the maintenance of patent applications up to the point of an issued patent. The justification for investments in TTOs is remarkably similar to that for university athletics, despite the fact that there is no clear avenue by which funds are returned to support the educational mission. Finally, just like in sports, there are only a few winners and the university's TTO is most likely to be one of the losers whose "team" does not do well.

Federal funding of research is still the mainstay of support for university research. The pressure to obtain federal funding is clearly communicated to professors, who spend an increasing percentage of their time writing proposals. Proposals for federal funding are submitted by scientists, or principal investigators (PIs), in response to specific calls related to a given research problem. The PI receives written critiques of the research along with a funding decision, which depends on the ranking of the proposal relative to the available funding level, known as the pay line. The pay line is the percentage of proposals that will be supported. The pay line can be as high as 30% in good economic times and less than 5% when funding levels are low, which means that the chance of receiving financial support in a given funding cycle has decreased from 1/3 in good times to 1/20 in recent years. The 1960s were the heyday for American scientific funding in proportion to the number of scientists. Today funding rates are at record lows in many categories [194].

It is difficult to calculate the waste in time and resources that is incurred when the funding rate drops to low levels as has happened in recent years. After 2000, government funding reached its maximum and has leveled off while the number of research active scientists has steadily increased. Prior to 2000 the NIH was in a unique position because of a mandated "doubling" of its funding level under the Clinton administration [195]. Nonetheless, inflation-adjusted budgets of other funding agencies had already begun to stagnate in the 1990s. The National Science Board 2010 publication entitled "Science and Engineering Indicators" cites a decrease of 7% in real dollars in federal funding since 2004, when governmental support for research in universities may have reached its peak level [196]. The environment in universities and medical schools today is working against accuracy of research reporting because professors and clinicians have less time to focus on the time-consuming job of checking results and looking over the notebooks of their students and post-docs [197].

When a funding decision is reached the agency enters into a contractual agreement with the university and provides the funding with certain stipulations. The university manages the funding and takes a percentage (28–69%) as overhead (indirect cost) and the PI uses the funding in accord with the budget approved by the agency to support personnel, supplies and equipment. Thus, the PI's traditional mission is to provide support for graduate students and post-docs who conduct most of the scientific research at universities. The inflation in the cost of education has had significant impact on available funding since tuition and salaries of graduate students (the research labor force in this enterprise) are the major expenses paid for by the federal grant funding system. Since the late 1980s, the academic inflation rate has far exceeded the general inflation rate in the economy, and is second only to the rate of medical inflation in its contribution to economic dislocations in the United States [198]. Scientific equipment and supplies are also subject to a particularly high rate of inflation, since universities are a captive audience of many of the companies offering scientific goods and support services. Scientists, medical researchers, and clinicians are disproportionately affected by these inflationary pressures as grant funding is increasingly consumed by rising costs. As the fraction of grant support that remains for the actual science decreases, scientists simultaneously feel the pressure to produce more in order to keep up with the expectations of program officers who increasingly want to have a flashy portfolio of research to show superiors at the funding agency and to Congress.

Overhead is not profit, but the dependence of the university on indirect cost receipts puts a pressure on university administrators that is similar to the pressure felt by management in a for-profit enterprise [199]. The administrators must position their institutions to compete for state and federal dollars in much the same way a company tries to increase its market share. The rise in number of competing institutions, and the success of large private research foundations, has kept an enormous pressure on institutions to upgrade infrastructure and hire high quality faculty. The need to hire well and push the professoriate towards more successful grantsmanship has put universities in a position where concern over the financial solvency of the enterprise often supersedes the appreciation of the scholarly pursuit of knowledge. While there have always been certain monetary incentives for those who applied for federal and state grant money, there is now a move in many universities to create a formal system that would calculate pay raises on a formula derived from success in raising grant money. This change in the system would put a new pressure on professors and remove part of their freedom to explore new directions, express their creativity in blue-sky research, or to provide constructive criticism.

State support of public universities has fallen for many years, in some cases leading to partial or even total privatization of their universities, in such states as Colorado, California and Michigan. For example, the University of Colorado (CU) now receives only 3% of its total support from state sources [2]. Consequently, universities in nearly every state in the U.S. have turned to their TTOs, private donors, and, of course, increased tuition to make up the difference due to the decreasing levels of state funding. While federal funding is still the most important, it has not kept up with inflation. Professors have been enticed by their administrations to solicit corporate research contracts and to patent their research to promote a new business model for the university. Since company and investor funding for university research implies private ownership of the products of research, the issue of who benefits from the university system becomes a central issue for society.

4.2 The Birth of Entrepreneurship in the University System

Early success stories based on academic science in the field of biotechnology helped to start an entire field of commercialization that has had a profound impact on universities. The first major patent in the new wave of activity was the recombinant DNA patent involving Dr. Stanley Cohen of Stanford University and Dr. Herbert Boyer of the University of California at San Francisco (UCSF) in 1975. Neils Reimers, a patent officer at Stanford University, first proposed the idea of a non-exclusive license to generate royalty revenue for the two universities [4]. While this new development defied decades of university policies shunning patenting for a profit motive, the time was ripe for a change. The economy in the mid-1970s had stagnated, but new technologies were emerging that had the potential to genuinely boost economic productivity. Indeed, the recombinant DNA patent generated $300 million in royalties and has become one of the most lucrative academic patents of all time. The birth of biotechnology as a technology transfer concept occurred almost overnight following the formation a company called Genentech by Dr. Boyer and venture capitalist, Robert Swanson. Of course, like any early stage company, Genentech took many years to reach its full potential for applications of recombinant DNA technology. Using host organisms, such bacteria or yeast human proteins such as somatostatin, insulin and human growth hormone could be produced in quantities sufficient for the marketplace.

Dr. Ivor Royton from the University of California San Diego (UCSD) founded Hybritech, using technology that enabled the mass production of antibodies for sale to research scientists for a wide variety of applications [200]. A few early biotechnology companies had a relatively short path to profitability since they were first in the market and this was particularly the case for so-called enabling technologies. Certain core technologies resulted in products that were able to produce a relatively timely revenue stream, as opposed the later arrivals, more specialized technologies, which carried much greater risk. Biotechnology companies based on the notion of developing medications from proteins or RNA produced relatively few blockbuster drugs initially and the path to profitability was long.

While Genentech and Hybritech have been extremely successful, the incubation times to profitability were 20 and 10 years, respectively. Many biotechnology companies that followed have had a more difficult path. The road to success in the treatment of disease is full of risks that are difficult to predict. Despite the obstacles and a trajectory quite different from traditional pharmaceutical companies, biotechnology has taken hold in a variety of areas of diagnostics, drug delivery, tissue replacement and gene therapy, in addition to the anticipated cloned proteins that would become the mainstay for large companies today such as Genentech, Amgen and Chiron [201]. Of course, companies such as Hewlett-Packard, Google, SAS (Statistical Analysis Systems) and many others in the hardware and software industries had definitive input from academic research. Since the founders started those companies after they left their positions as professors, the universities did not derive a long-term revenue stream from these companies. However, whether part of the university portfolio or not, most biotechnology companies lose money. The estimated total loss is well over $100 billion since the inception of biotechnology.

In 2013 a Brookings Institution study titled "University Start-Ups: Critical for Improving Technology Transfer" found that 84% of TTOs did not even pay their own costs [192]. Success is a moving target since patents expire [202]. Nonetheless, there are a few success stories. In a 2008 article in Forbes, Maureen Farrell describes how Stevens Institute of Technology of Hoboken, NJ topped the list in 2006 in terms of on-campus entrepreneurship in that year with a 16% rate of return based on investment in research, i.e. $4.5 million in patent income from a research outlay of $28 million. In that year Stanford received $62 million in patent revenue from a research expenditure of $700 million for a 9% return. Stevens Institute of Technology is a relatively small university that is named for Colonel John Stevens III, the founder of the Patent and Trademark Office, which may explain the focus on on-campus entrepreneurship.

The successes arising from patented university technology can be stunning. Perhaps the all-time success story in terms of patent revenue is the development of Remicade at New York University, which has generated billions in patent revenue for that university over the years. NYU Microbiology professor, Dr. Jan T. Vilcek led the team that developed Remicade, an antibody drug that suppresses the immune system in a controlled fashion for treatment of rheumatoid arthritis and related conditions [203]. It was licensed to Centacor and Johnson & Johnson. Aside from making nearly $2 billion in patent revenue, Dr. Vilcek donated $105 million to NYU Medical School [203]. Other historic high return academic investments include the pharmaceutical Somavert, developed at Ohio State University by the research group of Molecular Biology professor, Dr. John Kopchick, for control of growth hormone in the human body and marketed to Pfizer. Wake Forest University in Winston-Salem, NC has had high rates of return for medical devices and ranked second in Farrell's 2008 study [202]. Northwestern University Chemistry professor, Dr. Richard Silverman's group discovered the molecule pregabalin, which has since been licensed by Pfizer and marketed as Lyrica. Although plagued with side effects, Lyrica is an effective drug against epileptic seizures and neuropathic pain. It is an interesting case of serendipity in science. The original target was a neurotransmitter, but by luck it turned out that the molecule pregabalin could cross the blood-brain barrier, which gave it a range of functions not originally anticipated. The royalties totaling more than $1 billion have paid for a building on Northwestern University's campus among other benefits [204]. The revenue stream from the royalties for Lyrica was recently sold for $700 million, making Lyrica the highest revenue-generating drug from a university in 2008. In fact, prior to the marketing of Lyrica, the Illinois universities were considered lackluster in terms of technology transfer. However, the good fortune from this one blockbuster drug demonstrates why universities continue to pursue this financial strategy.

The success stories of university spin-off companies and patent royalties are the driving force for continued investment. Some of the technologies developed in universities have created jobs and helped to transform the economy into a more vibrant high technology marketplace. The Science Coalition Report documents 100 companies that can be regarded as success stories in the world of university spin-offs [205]. There is little doubt that if universities set out to embark on the road to riches through development of technology, there will be some big winners. However, there will also be many more losers who spend more money in patent and development costs than they recuperate. In a 2010 opinion in Forbes on-line Robert Langreth describes the successes and failure of biotechnology companies as equivalent to gambling in Las Vegas [206]. Thus, the path to commercialization of technology ties the fortunes of our educational system to the success of a few good ideas that may become drivers of technology. This trend in university research is coupled with massive investments of federal funding in certain target areas, such as healthcare

("The War on Cancer" or HIV research, for example), energy, and national security. The effect of this funding is evident even in the choice of words used to market research and describe progress in press releases. Words such as "fighting", "survivor" and "eradication" are used in a military metaphor for "attacking" disease. This use of language was not part of science prior to the political decisions to focus resources using a military analogy and therefore a mindset that was (and still is) foreign to most scientists. But the funding has a powerful effect on the ways of expressing scientific goals and ultimately even on the methods themselves. A war does not conjure an image of a dispassionate scientist carefully weighing the data in the balance, but rather a science with a mission who perhaps could overlook certain inconsistencies merely because they are deemed less important to the battle plan.

The problem with the investments being made in universities today is that it is difficult to predict a successful strategy. In almost every case, success has been the result of serendipity. It is not economically feasible for a university to simply patent everything although this does seem to have been a strategy followed by some. Even the technologies that have made it to the marketplace have taken decades to achieve success. Patenting hundreds of inventions over this period often ends up costing more than the expected payoff of the blockbuster drug. As if these issues are not already a concern for university priorities, in the midst of the rush by universities to develop a patent portfolio, the U.S. Congress passed the America Invents Act (AIA) in 2011. This new law changes the patent procedure from first to a First to Invent (FTI) to a First Inventor to File (FITF) system. This is a major change in patent law since it forces an inventor (or a university supporting the inventor) to file as quickly as possible. Under the prevailing law up to 2011 a carefully documented invention in laboratory notebooks could trump a patent filing if it could be shown that the invention was made prior to the filing. The FTI system in the U.S. was unique in the world, so those who supported the change point out that the U.S. has now aligned itself with the EU and other major patenting countries. But the U.S. system was working very well for the small inventor. Under FITF a small inventor must be nervous that all their hard work to conceive of a new idea and document carefully that it works could be scooped by another inventor who has deep pockets and can file quickly. The AIA is quite controversial in legal circles because of its potential conflict with article 1 of the U.S. Constitution [207]. The Constitution states "To promote the Progress of Science and useful Arts, by securing for limited Times to Authors and Inventors the exclusive Right to their respective Writings and Discoveries" [208]. This clause has been interpreted to protect individual inventorship. The implementation of a similar law in Canada resulted in a significant drop in patents granted to individual inventors [209]. A detailed article on the effect of the AIA in the biotechnology sector argues that there are strategies that can benefit small inventors [210], but one wonders whether the TTO in a typical state university will have the legal resources to take advantage of these. One obvious impact on universities is that the new law makes it even more difficult to share information since any hint of an idea may lead another person to realize the new possibility and quickly rush to patent it. Universities do not have deep pockets compared to investment firms and companies. A cash-strapped university TTO is at a significant disadvantage under the new FITF system. Since the law only took effect in 2013 it is still too soon to know if this law has killed the goose that lays the golden eggs for universities, but it has certainly dealt a major blow to their patenting strategy.

4.3 The Transition of the University Administration to Corporate Status

The relationship of corporations with university has been controversial, because of the potential for loss of academic freedom when the funding comes from a private sponsor. Over the past 20 years many large companies have reduced their in-house research and development efforts have come knocking on doors at campuses to find willing research partners who will conduct research on topics of interest to the company. Corporate sponsors can affect the university culture in other ways such as reducing the role of peer review, decreasing the free flow of information and independent rights to publish research findings. In the study “Big Oil Goes to College”, Jennifer Washburn documents the terms and conditions for ten major contracts from oil companies to university worth approximately $833 million [211]. In most of these relationships the university gave up the rights of academic partners to play any role in selecting the projects, evaluating the science or in peer review. Publication times were often delayed significantly. University research began to resemble contract research for an industry sector with a loss of core academic values, in spite of the fact that the goals of the oil companies were to find innovative ideas in alternative energy, a goal that many in academia would find desirable. Dan Greenberg points out that the economic results of the industry-academe partnerships have been mediocre for the corporations [2]. Many corporations have decided that their patenting strategy will be better served by pulling back and reinvesting in their own research and development. The percentage of industry funding has remained nearly constant for many years. Both Washburn and Greenberg agree that university patenting has financially benefitted a small fraction of universities [2, 211].

Regardless of the financial risks, universities are courting corporations with ever greater zeal in their constant search for new funding sources. Since Bayh-Dole the administrators see the synergy of small-scale patenting as well as the potential for patenting with a corporate partner. The requirements of federal funding sponsors have also shifted towards applied research. For example, nanoscience and nanotechnology are new areas for research that have resulted in shifted funding priorities. The emphasis in the case of nanotechnology started with “technology” and only later and less frequently does one hear references to nanoscience. The concept for the entire field of nanotechnology started at the highest levels of government with the intent of stimulating innovation in an area perceived to have great economic potential. Whether these applications have traction in terms of the development of products is another matter entirely. As with any top-down funding initiative a question that should have been asked initially is, “Is the number of scientists with nanotechnology expertise commensurate with the proposed amount of federal research funding?" What is certain is that given such a call for research proposals many scientists will apply, and they will all be using the appropriate key words.

The recognition of science as a business has lagged among scientists themselves even as universities have redefined their role as the locus for the commercialization of technology in the past 40 years. Scientists are caught the middle between their community values and the rapidly changing funding situation. The tradition in many disciplines is that fundamental research is prized, and this is also the training that many have received. But increasingly as they start their own careers they find that the money is in applied fields. As a society we are losing the benefit of the training in more fundamental fields and ending up with an increasing number of proposals in similar areas of research because "that's where the money is". The extremely competitive funding climate for federal grants combined with the encouragement of university administrators conspire to give scientists the message that they should contribute to the bottom line by submitting their ideas as invention disclosures. They can do this on their own, which requires great creativity and a leap into a new kind of thinking, or they can work with a corporate sponsor. The potential rewards are significant and the allure of being recognized by the university as one of the "producers" can play a role in tenure and promotion decisions along with the more traditional criteria of publishing and bringing in grant funding. These interactions produce multiple conflicts of interest that can lead to ethical conflicts. Most professors have no preparation for dealing with the business or ethical aspects of the development of technology for commercialization.

4.4 The Ethical Challenge Posed by Institutional Marketing of Science

Nanotechnology was conceived during the Clinton administration as a means of bolstering interest in basic science in a cross-cutting manner. The National Nanotechnology Initiative (NNI) mandated an increase of 1% annually in the funding from 2000 to 2010. While the original intent of the NNI started under the Clinton administration and was to obtain new funding for science over the decade from 2000 to 2010, the effect was to redistribute research dollars into the new field defined as nanotechnology. Since there was no new funding for science under the Bush administration, which started in 2001 [179], nanotechnology scientists benefitted, but many others who had solid research in non-privileged fields lost funding. Nanotechnology is an excellent example of marketing. Much of the nanotechnology research was not significantly different from previous research. For example, nanoparticles are actually colloids by another name. Colloids have been studied since the early nineteenth century. Colloids or nanoparticles, as they are now called, have severe constraints in their application due to lack of stability in liquid suspensions. The perception that there was value in nanoparticles led to investment and a great deal of new intellectual property in the nanotechnology domain. But some of the most widely touted applications of nanoparticles have their antecedents in immunoassays that have been known for decades prior to the advent of nanotechnology [212]. For example, the home pregnancy test has been known based on agglutination of proteins since the 1960s and analogous immunoassays based on gold nanoparticles were developed nearly a decade before the advent of nanotechnology [213]. However, the clever use of language in publication has hyped the field of nanotechnology. The slogans and catch phrases led to a common perception that nanotechnology is going to solve many technological problems. However, more than a decade since the advent of the NNI has revealed that nanotechnology is bound by the same difficulties as the rest of science. The commercial potential of the innovations may well have been oversold. A report by Cientifica emerged in 2006 entitled "Where Has My Money Gone? Government Nano-technology Funding and the $18 Billion Pair of Pants" documents the lack of return on a major investment of taxpayer dollars [214]. In hindsight, we can view the NNI as a nationwide experimental test of the Bayh-Dole act. Despite a great deal of marketing there have still been relatively few commercial successes tied to this initiative.

The attacks of September 11, 2001 resulted in a focus on homeland security. Threat detection, cybersecurity, rapid vaccine development and related security topics have suddenly become in high demand. Security research is an ideal area for spin-off companies as new markets develop as the war on terror continues to take new forms. Basic energy research has also seen a resurgence in part because of advances in alternative energies, but also because of the new technological innovations that permit extraction of shale oil and fracking for natural gas. Funding for medicine has supported major advances in vaccines for viruses from influenza to AIDS, Zika virus and Ebola virus. There are many spin-off companies in this area as the market for health care is enormous and prices of drugs and treatments has skyrocketed, providing a large economic incentive to innovate. None of this is lost on university administrators. They see opportunities to harness the technologies in their institutions that is justified with a new sense of urgency as state budget contributions continue to decrease across the United States.

Universities have enjoyed great advantages relative to businesses because of their tax-free status, and immunity from lawsuits of various kinds [215, 216]. But these advantages are beginning to be questioned as the business model for the universities is recognized as the same as that of other corporations. The universities that appear to be least affected by the growing pains during this period of transition are those with very large endowments. Endowments have become a central part of the financial strategy, even at state universities, since they provide the buffer that is needed to smooth the transition to a corporate structure. The 2010 General Accountability Office report documents that the rate of growth of endowments nearly doubled after 2003. Total endowments have increased from one hundred billion in 1989 to four hundred billion dollars in 2010 when adjusted for inflation [217]. Universities with large endowments typically make a return on investment that it is clearly a profit, so much so that the legislators in Massachusetts have studied the issue of taxing *annual revenue* from college endowments that exceed $1 billion [218]. The recent tax overhaul in the Trump administration introduced a tax on endowments for the first time. The $1 billion endowment club is growing by leaps and bounds as numerous state universities have surpassed that and limit others have set it as their minimum goal. As a consequence of the changes in the financial structure, the role of the university administration has evolved to include management of financial resources and strategy for product commercialization as important aspects of their jobs.

Recent studies such as the Goldwater Institute Policy Report have provided data showing a 61% increase in administrative costs during a period when instructional costs grew by only 39% [219–221]. One reason is that hiring of administrators has far outpaced faculty hiring. The increased private fund-raising efforts and technology transfer activities happen to have grown during this same period, which may explain some part of the growth. Administrative bloat has been so extreme in the state of North Carolina that the President of the 16-campus university system announced in 2009 that the 28% growth in administration in 5 years was "embarrassing" and required universities to cut administration first when facing new funding reductions [222]. The change in political climate has led to a different attitude in 2015 when the Board of Governors raised the salaries of the Chancellors of many of the state universities by record amounts citing the market for top administrators as a reason. One might wonder what happened to what used to be the primary mission of the university, which has traditionally been the education of students.

Administrators justify the increase in their own numbers and salary by size and complexity of the budget and management issues associated both endowments and technology transfer [223]. The increase in all financial aspects, the endowment, the patent income and associated spin-off companies, the increases in tuition, and the increase in certain specialized biomedical research funded project areas have conspired to increase the budgets of major state universities to the range of billions of dollars. The President of Purdue stated that the administration is actually very "lean" given the size of the $2.2 billion budget [223]. From all indications Purdue has attempted to keep tenure track professors as the mainstay of the teaching faculty. However, not all universities have done this. Many universities have hired low-paid "teaching techs" who can be recent graduates or sometimes students themselves to teach the classes. Administrator salaries have increased relative to those of the teaching and research professors as the goal of raising private donations and funding through technology companies has been deemed essential to the success of the university. It is ironic that many universities are cutting teaching staff and reducing many university services, while at the same time reporting record growth in the size of their endowments, and of course in the salaries of the administrators who are responsible for the fundraising effort. Recent studies indicate that the increased cost of the university administration has more than compensated for any savings achieved by these cuts in educational quality [224]. These trends truly cause one to ponder whether the focus of the corporate-university administration still is education.

The Bayh-Dole act is a Faustian bargain for state universities. As they increase their investment in patents and spin-off companies, public support is diminished by state legislatures and more recently even from the federal government [225, 226].

State legislators are increasingly seeing the price tag for higher education as an investment hoping for a good rate of return. The focus on the public good takes a back seat as legislators can rationalize that the free market will take care of the public need. Thus, the initial investment starts a feedback loop that forces universities to privatize further once they have taken the first step in that direction [115, 227]. Seeing the allure of being an economic engine and all that goes with that role many university administrators run the risk of losing sight of the educational mission. The effect on faculty is equally significant as their identity as scholars diminishes and their role in generation of technology changes the significance of the title of professor to that of founder or scientific officer in a fledgling company. The creativity focus has shifted to inventorship, which is now seen as an academic activity [228]. Academic studies suggest that greater public investment is needed for balance in the intellectual portfolio [229] and that the influence of commercial interests may weaken fundamental research [188]. To be sure there is the counter-argument commercialization will have positive outcomes for both research and technology transfer [230]. In the optimistic view, the role of universities as engines of economic development will transcend these negative aspects.

At a time when education has become for many the major investment of a lifetime, it is more imperative than ever for educators to ensure that the highest possible quality of public education. Despite these needs, the status of teaching at research universities has been on the decline for many years. Teaching is still an essential part of the job description, but counts for much less in faculty evaluations than any of the money-making activities that a professor can engage in. There has been increasing reliance on a cadre of non-tenured teaching professionals for service courses [231]. According to the American Association of University Professors the percentage of tenure-track faculty at American universities declined from 36.5% in 1975 to 21.3% in 2007, while the number of part-time teaching staff increased from 30.2 to 50.3% over the same period. This has created a two-tiered system in universities with great disparities in working conditions and academic freedoms [232]. Textbooks are a profit center that is now managed carefully by universities [233]. On-line resources [234] and distance education are offered to increase the revenue stream. The business model and entrepreneurship have extended to education itself with the for-profit universities such as University of Phoenix. The consequences of privatization of the university should be understood by the taxpaying public and by parents who trust universities to educate the next generation [235]. The net effect of the changes in management of the university in recent decades has led to an increased financial burden on the student and a decreased level of service. The responsibility for maintaining high standards and integrity becomes essential when business developers, investors, and corporate partners rely increasingly on university education and research [236, 237].

In today's complex world, examples abound of institutions that do not serve the average person particularly well. Financial institutions have become inscrutable for the average investor. Insurance companies manage healthcare in ways that are difficult for the average patient to comprehend. Universities are centers of learning that should help students find their way in a complex society. However, the trend towards increasingly private resources in the university means that the university administration is increasingly preoccupied with fund raising, corporate connections and investor relations rather than education [115]. The increase in financial pressure and opportunity provided by complexity have both resulted in numerous cases of falsification of research and even outright fraud at universities. Federal regulations require universities to investigate their own faculty if an allegation is made. Individual conflict of interest is recognized as crucial factor in decision making [238, 239], but there has been little recognition of the significant role played by university conflict of interest [240]. While the university administration may wish that the ethical hornet's nest would simply vanish, the temptation for faculty members to exaggerate and misrepresent their data for financial gain has never been greater.

Chapter 5
The Institutional Pressure to Become a Professor-Entrepreneur

It takes a thousand men to invent a telegraph, or a steam engine, or a phonograph, or a photograph, or a telephone or any other important thing – and the last man gets the credit and we forget the others. He added his little mite – that is all he did. These object lessons should teach us that ninety-nine parts of all things that proceed from the intellect are plagiarisms, pure and simple; and the lesson ought to make us modest. But nothing can do that.

– Mark Twain, Letter to Helen Keller, 1903

When we think of famous scientists, we envision thinkers who were driven by a deep desire to understand nature. The motive of financial gain appears to have been far from the minds of many of those who over the years pondered the nature of the earth, life upon it and the heavens surrounding it. Of course, many of the famous scientists prior to 1900 either had independent means of support or wealthy benefactors so that their relationship to funding was quite different than the conditions faced by scientists in a modern university. By all accounts, the financial aspect of the scientific profession was less important than the reward of being able to spend one's time in pursuit of discovery. Without a doubt, recognition for discoveries has been a crucial component of the scientific reward structure from the beginning. Sir Isaac Newton spent significant effort establishing that he, and not Robert Hooke should get the credit for codifying the laws of physics, and that it was he, rather than Leibniz, who invented calculus as a branch of mathematics. The truth is that each of these scientists made significant contributions, but Newton became a household name for the next 400 years because of his ego and insistence on taking credit for all of it. The traditional reward system in science has always been heavily weighted towards recognition and accolades from colleagues. While financial rewards of some kind have always been part of the mix, their relative importance has grown dramatically in the modern university.

S. Franzen, *University Responsibility for the Adjudication of Research Misconduct*, https://doi.org/10.1007/978-3-030-68063-3_5

5.1 The Changing Nature of the Reward System in Science

The modern university system with merit-based promotion according to a formula that includes significant increases in financial compensation is a relatively recent phenomenon. As a point of reference for lifestyles of famous professors in American history, J. Willard Gibbs is one of the most famous American scientists from the nineteenth century. Gibbs was a professor at Yale who never married and lived his entire life with his sister and her husband. Gibbs worked in almost complete isolation to systematize of the laws of thermodynamics in a way that modern scientists still find astonishing. He did so while always maintaining a low profile and a fastidious concern for mathematical accuracy and completeness. This example underscores the dramatic change that the scientific community has experienced in approximately one century, during which the university system began to institute a new formalized reward structure for scientists, known as the tenure system. In the current system, a scientist is hired as an assistant professor who has 6–10 years to establish a track record of productive funding and research, depending on the research field and institution. At the end of the probationary period, the faculty in the department vote on whether to grant tenure, which according to tradition provides job security for life. Upon conferral of tenure, the scientist is promoted to an associate professor [241]. Subsequently, after further publication and a demonstrated continuing productivity and funding a scientist may apply for the title of full professor. Each transition is associated with a substantial increase in salary. Along this career path, recognition may also come from awards form the federal government, from scientific societies and from private foundations.

In the Western world, one can view the prizes for discoveries given by the Royal Society as an early example of the type of academic prizes that exists today. However, one would be unlikely to be able to count on the Royal Society of the eighteenth century for objective peer review as discussed in the story of John Harrison in the book *Longitude* by Dava Sobel [242]. In 1714 the British Parliament passed the Longitude Act, which provided a prize of 20,000 British Pounds for the invention of a device that could predict the longitude with an accuracy of 30 miles for a journey from Britain across the Atlantic to the West Indies. A clockmaker and commoner, John Harrison solved the problem by the invention of an accurate and stable marine chronometer. His invention was the clear winner of the contest proposed by the Royal Society. The fact that Harrison lacked a university education, as well as the fact that he was not of noble birth, prevented the Royal Society from accepting its charge to award him with the prize until very late in his life. Sobel's book discusses how it required several acts of Parliament to override the decision-making process of the Royal Society and award the prize to Harrison. Today there are hundreds of prizes for achievement in academics and it is common for professors to garner support from colleagues to be competitive in the struggle to establish one's value to the university. One could argue that winning scientific prizes is as political today as it was in Harrison's day since it involves value judgements about the relative merits of scientific contributions.

In the century since Planck's and Einstein's seminal work that ushered in the quantum "revolution" (a la Kuhn) the scientific enterprise has been transformed [119]. In post-world-war II universities there has been an acceleration of the trend towards scientific conformity with a standard that mandates one should devote oneself as completely as possible to one's specialization, which becomes the identifier for each scientist. The mechanisms to promote the field and one's place in it are not a set of written rules. Rather, scientists learn the unwritten credo of academic standards in graduate school. The senior colleagues propagate these behaviors to groom faculty for success in the tenure process. Those who do not play by the rules are weeded out. The goal of the publish-or-perish system is to force each professor to identify a niche of science that can be used to identify him or her. Failure to clearly define a unique project and obtain funding is grounds for refusal of tenure. For many years, the idea that academic scientists would work for profit was considered anathema. Scientists relied on their institutional salary and the focus on research precluded earning extra income. However, that view of research scientists has been replaced by a new one, in which scientists can have a much broader impact and they can reap financial rewards commensurate with their contribution. Neither the professoriate nor university administrators have a clear idea how this new mandate should relate to the traditional value system of academia.

The increasing reliance of society on technology has made certain careers in science into positions that can have an impact on industry and government. The expectation in research universities is that the professors in science and engineering will bring in funds that are greater than their annual salary (perhaps many times greater) and thereby contribute to the bottom line of the university. With the emergence of pressures to develop a new funding model based on patenting and entrepreneurship, universities offer another potential avenue for professors to use their talents. The role played by journals in the promotion of science, the marketing of technology developed in universities by TTOs, and ultimately, the money that can be made from those activities have all conspired to reinforce the tendency to market the professors themselves as a product of the new corporate university structure. With this new emphasis comes the possibility that professors can become stars based on their success in fund raising, separate from any consideration of their contribution to the fundamental processes of scientific discovery.

5.2 Academic Stars and Perverse Incentives

Albert Einstein's career ushered in an era, in which a scientist could have celebrity status. Although recognized as a bright mind as a young man, Einstein was invited to join academia only after his seminal papers on the theory of relativity were published, while he worked as a patent clerk. The idea of a scientist publishing some papers while working outside the university to gain entry into the halls of academia is virtually unthinkable today. In the modern university, hundreds of aspiring candidates compete for each academic position. Only those who have obtained a PhD and one or even two post-doctoral positions will have a chance of receiving an offer of a job as an assistant professor in science in today's university. Moreover, priority in hiring is often given to those who studied in research groups with "pedigree" meaning that they are at top universities and worked with a mentor who is at the top of his or her field. As the competition for external funding and entrepreneurship have increased in importance, there is a new kind of star that also brings in large sums of money into the university. While it is normally assumed that scientific achievement goes hand-in-hand with funding there is no requirement that they be correlated with insight or scientific scholarship in the traditional sense. Academic stardom today is celebrity status within the university administration, which rewards those who are producers for the coffers of the university. Sadly, given the limited resources available for scientific research the changes in the funding model have occurred at the expense of others who follow a more traditional career path focused exclusively on scholarship and teaching. Certain "hot" areas such as nanotechnology have arisen at the expense of funding in established categories. These trends have increased the pressure in academia, both the pressure to make it as a star and the pressure to survive in a world of shifting funding priorities.

Today there are scientists who appear to have inordinate success with publishing, patenting, and leading new funding initiatives. Their success in funding promulgates a myth of scientific exceptionalism. In reality, the enormous productivity derives from a large research group that can easily be 30 or more talented graduate students and post-docs. Of course, the PI must be a master at writing scientific publications and conference presentations, but also a skilled manager able to delegate to a large diverse group of talented people. A person who succeeds in combining these skills can have an exceptional career, which is reinforced by trends in the funding mechanisms. There has been a tendency to concentrate research funding into larger sums, that is into center grants or multi-investigator grants at the expense of the smaller single-investigator funding mechanisms. The "star" scientist is poised to obtain these larger grants that others simply cannot compete for. Often, a "star" professor is identified by the university administration and funding agencies based on the perceived value of their ideas and that perception is sometimes influenced by their apparent marketability. Once "stars" have had a certain measure of success in fund raising, the associated reputation helps to obtain center funding and other large multi-investigator grants. Program officers like to identify winners since that makes it easier to make funding decisions. But star quality in science is not the same thing as deep scientific insight. Even when the goals are admittedly financial, the horizon for a tangible result is often sufficiently long that one may not know whether a particular investment in a line of research is worthwhile or not. The system is moving in a direction of placing fewer bets on new science and technology. Moreover, it does this in a way that sometimes favors marketing over substance. If a PI in charge of large grants is highly competent and energetic about making sure that the team effort is working well, this could be an efficient funding model. However, if the PI is not strongly interested in the management of the grant and coordinating the team effort, there will likely be waste.

Merton noted that one feature of the reward system of science is a tendency for those who are established and recognized to receive further recognition [243]. Merton called this "the Matthew effect" after the biblical passage from Matthew known as the law of abundance, "For unto everyone that has shall be given, and he shall have abundance: but from him that hath not shall be taken away even that which he hath." The Matthew effect can be seen in the observation that the recognized scientist often receives credit for ideas of others in a research group or on a publication. One way that scientists confer credit is by eponymy, the naming of a theory, effect or apparatus after the scientist who made the principal contribution. One could say that Newton's laws are the quintessential example of name recognition since they are the laws of classical physics. Hooke got second prize with his name on the physics of vibrations, specifically oscillating springs, a particular subset of motion, but not the general laws that Newton would lay claim to. Harrison, the inventor of the marine chronometer, has received less acknowledgment than more theoretically inclined colleagues. His invention was never called the "Harrison marine chronometer", perhaps because of his lowly status as a mere clockmaker. However, in a practical sense, one could make the case that Harrison saved many more lives than any scientist of his era. Shipwrecks were common prior to Harrison's invention, due to lack of ability to accurately measure Greenwich Mean Time on ships. Today, name recognition is easy to establish through publication, but also increasingly through invention, patenting and entrepreneurship.

Merton concluded that there is a strong tendency in human behavior to devalue the contribution of less known co-authors on a paper and to remember, as well as recognize the academically established figure as the originator of the work. In many cases, the students who carried out the work and even have contributed significant insights are not widely recognized. Instead, the name recognition of their mentor carries the weight in the academic community. The Matthew effect is as important today as it was when Merton first coined the term, but it is further enhanced by the success of the contemporary science-entrepreneur who brings in major funding to the university. It is as though the Matthew effect today extends to the financial reward system, making the name of the scientist a brand that inspires confidence in the program officer or investor. The criteria for recognition as an academic star include building a large research group, obtaining large grants and publishing. Winning awards and obtaining the support of the administration are also essential. But why concentrate resources into a small number of large research groups rather than distribute them to the largest number possible? The current model for scientific recognition has failed in three ways: 1. Those doing the evaluation frequently do not understand the science outside of their limited field, 2. Those called upon to render a judgement are themselves very busy and cannot take the time to delve into the complexities that might have to be overcome to make a truly valid evaluation, 3. There are widely different interpretations between fields, and sometimes even within fields, as to what constitutes an important research area. On the other hand, the current funding model is susceptible to self-aggrandizement. In many branches of science, the evaluation of academic scientists has inherently come to include patenting and commercialization of technology as an important component. Filing a patent has many of the same issues that peer review does. Patent examiners may reject a patent application for prior art, but the complexity of science makes it more difficult to reject a patent for lack of feasibility. By the same token, starting a company does not in itself mean that the technological foundation is necessarily innovative or sound. In fact, significant marketing is often necessary to help others to recognize the validity or applicability of even the most solid idea. Thus, the new avenues for obtaining recognition and funding are dependent on self-promotion to an even greater degree than government funding. The dramatic increase in the number of scientific journals has changed the landscape for publishing to include many more practical or potentially commercial innovations as well. These changes have occurred so rapidly that many scientists have little context for evaluation. These considerations clash with the traditional values of scholarship.

How does the university administration recognize an academic "star"? Once a scientist has a certain measure of success in the funding arena, perhaps complemented by patents and other financial success, the university administration is likely to choose that scientist to lead other larger efforts to garner funding. There will often be an internal competition for a competitive position as the PI to lead a multi-institution grant or a center grant. The internal selection criteria will usually be based on publication track record (often measured as the h-index) and general funding track record. While a university administrator can quantify total grant funding and success in commercialization in terms of revenue generated, there is no true metric for the quality of a scientific discovery. Dr. Marc Edwards of Virginia Institute of Technology has argued that the incentives in today's university are "perverse" [5]. Since everyone knows what numbers need to be maximized to evaluate a scientist, there are many ways in which an individual can increase their apparent worth to the institution. Publication is the currency used to justify grant funding. Although many academics do not want to openly admit that maximizing the number of publications is crucial for their success, the closed-door meetings to determine promotion, and conversations in panel review to determine funding often mention the number and quality of publications as a crucial parameter to advocate for a positive result. To help with their own success, scientists can suggest each other as reviewers, nominate each other for awards, and cite each other's work to enhance their standing in ways that have nothing to do with scientific accomplishment. The metrics employed by the university administration have little to do with being a good scientist or a team player who will actually bring scientists together to develop the type of synergy that the funding agencies are trying to foster. All too often, the center grant funding turns into a large pie divided by each participant who uses his/her slices as needed for project goals or other more pressing needs. If the administrators and funding agencies have recognized true talent, then the academic star's ideas will pay off in the long run. The time horizon for science is often quite long. By making fewer "bets" on new ideas, the system maybe lessening the probability of future payoffs.

Scientific citation has become perverted as measure of merit by the very fact that everyone knows that in the modern academic career quantity trumps quality, although the reverse should be the case. Gardner and Veliz studied tenure policies and outcomes in a "striving university" to understand the effects of academic pressures over three decades [244]. Despite the "striving university's" declaration it would prioritize quality over quantity at the start of the study period, "we note that none of the units in 1980 had any detailed guidelines or specification for the *amount* of scholarship to be produced, but 12 of them did by 2012." The most common example of such a quantitative requirement is a fixed number of publications required for conferral of tenure. The sheer number of publications cited in any grant or promotion package is prohibitive of any detailed analysis for quality. Quantity is takes precedence over quality. To increase one's valuation it is possible to make deals with other scientists. Scientists can agree to review each other's work and to assist each other in building up their citations. To a certain extent this is normal, and it is the kind of thing that happens at conferences and visits to each other's institutions to present seminars. By meeting and discussing science there is a natural tendency to cite each other's work and scientists in the same discipline who know each other will be likely to see each other's papers and grants in the review process. However, the underlying need to promote certain metrics in order to compete in the arena with others for the "star" position as PI on large grants gives scientists a reason to worry more about catering to reviews and citations than being primarily concerned about the underlying quality of the research. The perverse incentives cause scientists to spend more of their time on a certain kind of social jockeying for position. Activities that should be irrelevant have taken on an inflated importance because publication and funding determined by quantity rather than quality.

Scientists build up their "brand" as pioneers in their field and work on getting their name recognized for their accomplishments. If they can appeal to a contemporary community, they can build their publication and citation record. There is an incentive to make one's work appear to be independent of prior work in the field. A failure to appropriately acknowledge prior work is antithetical to scientific traditions, but the academic literature is so vast that it is possible for a scientist to move aggressively into a field and fail to cite prior related work. This may happen out of genuine ignorance, but it can also occur out of self-interest. Self-aggrandizing behavior would have been counter-productive a few decades ago when the economic pressures were not as great and the funding was more geared towards hard-nosed evaluation of the proposed science [120, 245–248]. The weakness of decision making in universities is manifest in tenure decisions for faculty, in which the number of publications is second only to the number of dollars brought into the university in determining eligibility for tenure and promotion. As funding becomes more competitive this model is fraught with contradictions and a great deal of pain for those who are starting academic careers today [194]. For those who aspire to do great research, but not to stardom, the avenues for success are fewer than in past decades. The funding has favored the larger grant mechanisms and the administrations have doted on those who bring in the big money.

There is a kind of futures market in scientific innovation that plays out in the TTOs and in the perception of the administration regarding which technology is of potential value to the institution. As is the case for complex financial instruments, the true value of an invention or discovery may be difficult to evaluate because of the complexity of science. Just as in the business world, there is a spin one can put on achievements that can make them seem more significant than competing ideas and perhaps more feasible than they will turn out to be. Some scientists have a reputation for getting results. Since results are often measured in terms of publications, an academic star pushes out publications and specifically targets high impact or luxury journals.

Given the pressures associated with the star phenomenon one must ask whether it is a coincidence that many of the famous cases of scientific fraud in recent years have involved publications in the highly prestigious journals *Science* or *Nature*? A recent study suggests that a star is much more likely to be able to get away with fraud [249]. In the interest of highly polished presentations it is possible to engage in questionable practices such as the omission of certain details in the method or particular data that are inconvenient. Scientists admit that such questionable practices are prevalent [250]. This does not mean that the stars at the top of the pyramid are less ethical, only that if they choose to be unethical, it will be difficult for others in the community to recognize it, let alone say anything in cases where questionable practices are detected.

5.3 The Consequences of Promoting Professorial Patenting

Because of the Bayh-Dole act professors who invent new technology have the potential to generate revenue for the university. In return, the universities have given incentives to research-active professors, by giving them a share of any profits derived from the sale or licensing of intellectual property. The precise division of royalties from intellectual property and the ownership of patents is set by the administration at each university [251]. A common arrangement is for the university and the inventor to split profit for patent licensing or royalty income 50:50. The reward for patenting in universities is typically quite generous compared to the terms employees receive in private industry. While the university owns the patent, just as a corporation would, university professors have a much greater incentive to patent their science because of the excellent terms given by universities [229, 252]. In a typical private company, the inventor may receive a one-year bonus, but no share in patent licensing or royalties. One famous example of patent compensation in private industry is that of Kary Mullis, one of the inventors of the polymerase chain reaction (PCR), which permits amplification of specific sequences of DNA [253]. For conceiving of a patent worth billions of dollars, he received a ~ $10,000 bonus the year that he discovered the technology while working at Cetus Corporation. Mullis was later awarded the Nobel Prize for his invention of the technology, but most inventors in industry do not win any prize for their inventions.

The allure of success in biotechnology and nanotechnology has led many professors to patent their technology and look for business development teams to help them start a company. This type of activity is precisely what was envisioned by the Bayh-Dole Act and is now encouraged, and strongly supported financially, by the university administrations, and by considerable federal funding through the SBIR (Small Business Innovation Research) and STTR (Small Business Technology Transfer) program and contract funding mechanisms [120]. The path to commercialization exerts great unforeseen pressures on the university professor who is the source of the ideas for this new industry [254]. The 2013 Brookings Institution study acknowledges the political pressure for universities to move in the direction of increased output in the form of start-up companies [192]. The study also states the plain fact that the results so far are disappointing. Now that the model of an economic engine is clear, society is waiting for the academic stars to come up with solutions and get companies started. But does it work like that? Is there a formula for success that involves a sprinkle of STTR funding, a nurturing start-up incubator and a supportive seed funding program that will make the start-ups flourish?

There is a culture clash in the attempt to involve a faculty member in the process of commercialization [255]. The expectations of commercialization teams on one side, are at odds with methods and rigor of academic scientists on the other. Simply put, good science is not compatible with rapid commercialization. The commercialization team wants to develop a product starting from the point that the technology is on sound footing. The scientist would like to garner support for an idea and is often working frantically to show that the technology is ready for commercialization. Due to pressures on both sides to obtain a result, there is ample room for miscommunication about what is expected by the commercialization and the readiness of the technology by the academic researcher. Investors may prefer that the TTO weed out the immature technologies, but it is not always obvious, which technologies are appropriate for further development. In this regard the inexperience of academic technologists is viewed as a major impediment by the investment community. An objective long-term vision that includes both the business and technology perspective is needed to really take advantage of what technology has to offer [254]. Thus, the development of ideas that will be ripe for technology transfer is a difficult matter that is often poorly understood by the technologist (professor) and the commercialization consultant who seeks to use the idea as the basis for a commercial enterprise [256]. The passage of the AIA, the first-to-file patent law, which came into effect in 2013 has complicated these considerations and greatly increased the pressure on university TTOs to patent as quickly as possible. This law is likely to exacerbate the pressures felt by the TTO and even more so the professors who are developing technology. The implications for ethics are obvious. With less time to check and verify, there is a high probability of rushing to patent before all experimental work is completed in the fear that delay will result in a loss of the patent.

As a result of the incentives to faculty, the modern university has become an incubator for marketable technology. For example, research into anti-cancer drugs or chemotherapeutics has attracted a great deal of attention as an application in chemistry departments. In reality, most of these research efforts are quite far from clinical applications, but they are sufficiently relevant to the medical condition that chemists can reasonably expect a favorable outcome for a proposal to the NIH or a patent application. There are many stipulations that must be met before a drug is ready even for first phase clinical trials. Some of the most successful anti-cancer drugs, camptothecin, and taxol, both required nearly 30 years of research before acceptance for clinical use [257, 258]. These drugs were discovered from natural sources by screening methodologies. Today, most drugs are developed de novo, which requires years of prior work before the clinical stage. Because of this long timeline, both scientists and investors often lack the patience needed for development and commercialization of new therapies.

It is a laudable goal to want to cure cancer and many research scientists take up the challenge. One reason that academic research groups move in this direction is that cancer is less difficult as a scientific target than many other kinds of pharmaceutical research. The nature of cancer research in its initial stages is particularly tractable for non-medical research scientists, since cancer cell lines are very robust and easier to maintain in the laboratory than normal cells. The first step of the drug development process has a relatively easily defined goal: to kill the cancer cells. Nonetheless, killing cancer cells in a Petri dish is only the first in a very long series of steps that are necessary to develop a chemotherapy drug. There are thousands of failed chemotherapy agents that will likely never be used. Behind each of these drugs are years of research and major financial support to get the drug to the point where it could be evaluated [259]. Many of these drugs have arisen as a result of academic research projects at university medical schools, in chemistry or medicinal chemistry departments. Since the "War on Cancer" declared in the 1970s by then President Nixon, a great deal of grant funding has been devoted to this type of research [260]. The fact that professors and academic researchers can become wealthy in the process may be considered just another facet of the entrepreneurial society. However, we do need to be concerned about how resources are allocated and scientific claims regarding the fight against cancer need to be carefully justified. The complexity of biomedical research gives rise to a minefield of potential conflicts of interest.

The history of laetrile provides insight into the ethical issues confronted in this type of research, so commonly undertaken at universities. With a 120-year history, laetrile is one of the most famous fraudulent anti-cancer drugs ever known. Laetrile has given false hope to many who wanted to believe in its efficacy, despite the overwhelming scientific evidence that it has no effect. In fact, laetrile was legal for a long time in some other countries and it was sold for profit to patients who would travel to those destinations to take it. However, laetrile has been thoroughly investigated, and there is agreement among scientists about its lack of significant effect. Ideas that have been generally accepted as fraudulent seem to be revived repeatedly by fraudsters. This is why it is so important to prevent fraudulent claims from being published in the first place. Laetrile was tested for its anti-cancer properties in Germany in 1892 and rejected as ineffective. However, this rejection did not prevent its continued use and eventual resurgence in the 1960s [261] until, based on the clinical evidence, the Food and Drug Administration banned routine use of laetrile in the U.S. in 1965. Despite the published studies and the ban there was still a popular belief among terminally ill cancer patients that laetrile would be effective. In 1977 this issue reached a fever pitch when several states legalized the use of laetrile for terminally ill patients [262]. On humanitarian grounds it is hard to deny a dying person their wish and this appears to have been the consideration in a court case permitting terminally ill patients to import laetrile, which was banned in the U.S. (United States vs. Rutherford, 442 U.S. 544 (1979)). However, clinical trials run in the U.S. in 1982 showed once again that laetrile had no effect against cancer [263]. By this time laetrile had many proponents in the public who have believed in its efficacy and even claimed that there has been a conspiracy to block access of patients to this "beneficial" treatment. However, none of this changes the assessment based on the scientific data. While laetrile is so far on the fringe of mainstream medical research that its use can be regarded as quackery, the issues surrounding the acceptance of even mainstream pharmaceuticals has been contentious both from a scientific and ethical perspective.

It is precisely because of drugs like laetrile that we need to understand the dangers of ethical missteps in the development of chemotherapy agents, and indeed any pharmaceutical that purports to treat a human disease. In some instances, companies have been started at universities and medical schools before a drug is proven based on claims made by researchers. Since private investment is usually needed to pay for clinical trials the goal of early stage academic researchers is to obtain results that are sufficiently compelling to warrant the investment. Ethical issues abound in this process since the promotion of an idea, the marketing of the science to the granting agency or an investor support are all based on the understanding and presentation by the professor. Because financial incentive is a significant part of the drive to discover new medications, researchers must ask themselves difficult questions from the earliest possible stage, and keep asking them, to avoid self-deception or unethical treatment of subjects. National Public Radio journalist Richard Harris recently wrote the book "*Rigor Mortis: How Sloppy Science Creates Worthless Cures, Crushes Hope, and Wastes Billions*", whose title leaves little to the imagination [197]. The book examines the waste in the result of investment of NIH funds for the development of new cures and medications. Many of the issues in clinical trials are the result of poor planning or lack of expertise, but clearly ethical aspects are also a great concern in medical research.

The sums of money spent on the commercialization of university research are far from trivial. One recently publicized example illustrates the nature of the issues and the magnitude of sums of money that can be involved in the sale of technology developed at a university. In 2003, Dr. David Sinclair of Harvard Medical School reported that resveratrol, a compound found in wine, could prolong lifespan by up to 70% in nematodes [264]. Resveratrol has been known since 1938 and has been implicated as a candidate molecule that may explain the "French paradox", namely that incidence of heart disease is relatively low in a population that eats rich food and drinks more than average quantities of wine. In 2006, even more exciting results were published in *Nature* that showed that the lifespan of overweight mice could be prolonged by up to 50% [265]. The company Sirtris was founded based on the technology in these articles. In 2008, Sirtris was purchased by GlaxoSmithKline (GSK) for $728 million. In 2010 GSK voluntarily ended a clinical trial involving resveratrol because it was found not to have the efficacy initially expected. While these events may be unrelated it signifies that decisions regarding the value of resveratrol and related drugs had to be reconsidered. The most significant scientific issue came to light in 2011, when it was determined that the studies published in *Nature* 2006 could be not be reproduced [266]. The findings of Burnett et al. in 2011 were not simply a challenge to resveratrol, but rather to the entire notion that sirtuins play a role in the aging process. The issue of the role of sirtuins has become highly controversial with several members of the original researcher team disputing the significance of the findings [267]. While there has been no public statement regarding the technology and Sirtris continues to work on several drugs in clinical trials, the facts suggest that an optimistic interpretation of academic research on at least two fronts was the basis for the valuation of the technology by GSK.

The spectacular failure of the proposed anti-dementia drug Dimebon (latrepiridine) is another example of misplaced optimism based on academic research. Pfizer purchased the rights to an anti-dementia drug, Dimebon, for a total of $725 million. Dimebon is an anti-histamine discovered originally in Russia. In 2008 a group at Baylor College of Medicine reported a clinical study using Dimebon as an anti-dementia agent [268]. The data appeared sufficiently convincing that commercialization was planned. Pfizer tested the drug in clinical trials with a company called Medivation, but the drug failed to show efficacy in phase III clinical trials [269]. While this may be a case of simple bad luck, the fact remains that the lack of reproducibility of university research (see Sect. 4.8) combined with the pressures from the TTO to market research results has the potential to lead to poor allocation of resources. These examples suggest a need for greater caution in the interpretation of results before startups are formed and drugs are marketed to large companies.

5.4 Crossing the Commercialization Desert: How Many Academic Startups Lose Their Way?

Commercialization requires crossing a desert between proof of principle and the marketplace. Obtaining the funding to sustain an effort to develop a new technology presents unique set of ethical challenges. Miscommunication regarding the readiness of the proof-of-principle technology can introduce time pressure that has unintended consequences. The expectation that the technology will be implemented rapidly can force researchers into an awkward position where they are trying to get a certain result by a designated date to reach a milestone. Short cuts are not necessarily unethical, but the outcome may lead to a compromised result that is presented to an investor or a patent attorney before it is completely understood. The outcome may be that later it becomes clear that the result is not scientifically sound. Thus, a researcher may be required to make a hard choice between admitting an inability to produce the desired result or trying to make the data in hand appear to be sufficient to meet the milestone.

There are various strategies for the endpoint of a spin-off company. Many university companies lack the resources and experience to reach the marketplace and therefore are hoping to be bought out once they have established a sufficient proof of principle. The climate in the university provides a temptation to push ahead with an idea that one knows has technological deficiencies in the hope that these will be overcome. Many of the ideas never make it to the product stage so the fact that significant resources may have been wasted is not appreciated until later. The inventors may still be highly successful based on university metrics since they may have garnered research support and consultancies even when the idea that was proposed fails completely. As with biotechnology in previous decades, the payoff for nanotechnology research has a long horizon. Nanotechnology came onto the scene as the dot-com bubble was reaching its capacity for expansion. Investors and established major companies were flush with cash. There was a great deal of "shopping" in universities for new ideas and start-up companies. In the late 1990s the risk of purchasing a small company must have seemed acceptable, since many were purchased, even though these transactions delivered little or no value in the end.

One of the central promises of nanotechnology was to further miniaturize electron circuits based on arrangement of individual molecules as components of a circuit. This strategy goes by the name molecular electronics. A company with the name Molecular Electronics was formed by two professors, Dr. Mark Reed and Dr. Jim Tour in 1999. The idea was based on a publication in *Science* that reported that a single molecule could conduct electricity [270]. The result reported in *Science* formed the basis for a concept that molecular scale (very tiny) transistors could be produced [271]. Although an undisclosed number of millions of dollars was poured into the company it yielded no products and closed its doors within a few years. Research in the area has been plagued by lack of reproducibility and has remained an academic pursuit more than 14 years after the report of a breakthrough [272]. The present issue is not whether the technology has long term prospects [273], but whether it was oversold at an early stage. The academic and entrepreneurial cultural mix provides a new set of challenges for scientific ethics.

5.5 Commercialization Guided by Hype Can Lead to Fraud

Marketing and patent strategy put time pressure on a scientist that changes the meaning of scientific publication. Rather than full disclosure of ideas that can be replicated, which is a standard for most scientific publication, the entrepreneurial scientist may need a validation that simultaneously does not release too much information. The spate of publications of "one-hit wonders" in *Science* magazine has increased in recent years. While it is easy to simply document the number of researchers who publish only one time in *Nature* or *Science* [274], the more pertinent use of the term "one-hit wonder" refers to papers that make remarkable claims with little or no follow-up. While publications of dramatic new findings would normally be followed up with documentation of the background experiments and characterization that was done, there has been a tendency to rely on publications of results that often are never supported by a detailed analysis. Yet, having a publication in one of the luxury journals is often sufficient support to patent and promote the concept as a technology through a university TTO.

The path to commercialization pushes marketing of science into a realm with which most scientists are not familiar. There is an uneasy balance between a business development technologist who wants to shape the message and an academic research technologist who needs to ensure that the message is consistent with all the facts. The scientist must, to be true to his/her calling, be careful to be completely accurate and not embellish the scientific message for marketing purposes. The business developer, who often does not fully understand the science, must understand the university setting and the pace of research to accurately portray the state of the technology and timeline to would-be investors. Problems arise when there is an expectation that technology can make greater strides than is realistic. The added pressures result in greater haste and therefore less time to carefully consider possible experimental "artifacts", which may be mistakes or, perhaps by serendipity, indicative of a novel result [120]. The education of the professoriate in the potential pitfalls of entrepreneurship has not kept pace with the excitement associated with the potential for patents and spin-off companies in academic circles [275]. Recognition of the associated responsibilities of the university administration has lagged behind as the portfolio of patents and spin-off companies propels them into a new kind of financial activity.

In 2015 a series of articles in the Wall Street Journal documented the questionable business practices of a rising star in the corporate world. The company Theranos was founded by a former graduate student from Stanford University, Elizabeth Holmes, with support from university mentors, several of whom had connections with other known biotechnology companies with roots at Stanford. The technology was kept secret and so it is still not clear what the innovation was. The company claimed that a new more effective blood test had been developed that used a finger prick rather than the standard 5–20 milliliters of blood withdrawn by a needle from a patient's arm. The proprietary tests competed with established companies such as LabCorp for the routine clinical testing market. At the time of the Wall Street Journal articles, Theranos had investments of $400 million and a valuation of $9 billion. The CEO, Elizabeth Holmes, dressed in signature black turtlenecks, was seen as a new Steve Jobs style entrepreneur in the health sector [276]. There were significant technical problems with the technology, which were evident from the fact that company was forced to abandon its proprietary assays since they failed to give accurate results in all but a few instances. For some time Theranos had been advertising the use of the new methods while using standard technology for most of the tests it conducted. At the time of the first article in the Wall Street Journal the FDA made a surprise inspection of a Theranos facility and found that only 1 of 240 tests actually used the proprietary technology. The remaining 239 tests were carried out using standard technology that the other major corporations were using. Of course, it is ethical to compete using the same technology, but it is not ethical to misrepresent the technology that you are using. Theranos gradually lost all credibility and federal regulators revoked its license. A criminal investigation was launched subsequently into the claims made to investors [277]. Lawsuits by investors and patients who claimed that they suffered harm from incorrect information due to the inaccurate blood tests followed [278]. Today, Elizabeth Holmes is facing allegations of significant fraud that may ultimately lead to prison time. In this instance, the company was not started by an academic, but Ms. Holmes surrounded herself with Stanford professors who were willing to serve on the Board of Directors or otherwise give credibility to the fraudulent company.

Scientist-entrepreneurs are common in universities today. But the transition of universities from Ivory Tower to commercialization hub transpired so quickly that it is difficult to find individuals who can bridge the chasm between the worlds of science and business. Scientific training is poor preparation for the world of business. Whether intentional or not, certain ideas have been hyped and sold to investors even though the technology was not ready or perhaps even faulty. The interdisciplinary nature of research has increased the complexity, which has increased the number of ways that scientists can deceive themselves, as well as others. Experienced scientists have noted that it is harder than ever before to recognize genuine advances from hype [279]. It is reasonable to hypothesize that this tendency has grown as a result of the marketing of science, both as the sale of intellectual property to investors and as the promotion of solutions to program officers in government funding agencies. Just as has happened in financial bubbles where real estate and derivatives were sold at inflated prices, the expectation that new technology will produce breakthroughs in the fields of homeland security, energy and medicine, among others, has created a milieu in which scientific ideas have been sold without the requirement that the fundamentals of the technology have been thoroughly understood.

Chapter 6
The Short Path from Wishful Thinking to Scientific Fraud

> *If [Newton's] Principia established the quantitative pattern of modern science, it equally suggested a less sublime truth – that no one can manipulate the fudge factor quite so effectively as the master mathematician himself*
>
> *– Eric Westfall, Science, Vol. 179, p. 751, 1973*

While the traditional emphasis on fundamental discovery forms the basis of science, the connection to real world applications and problems has become an essential ingredient to obtain funding. As specialization has increased, it has become increasingly difficult to judge scientific achievement, which makes communication of the significance of one's work all the more essential. The funding agencies are quite direct about requesting flashy advertisements of research accomplishments to show Congress and other governmental observers. Scientists are under pressure to promote their science in every way they can. This means learning how to write grant proposals to make a case that the research has a benefit to human health, to the environment, or to the economy. Corporate funding requires scientists to adapt their effort to the needs of the external sponsor. Even the most careful and honest scientist will feel some pressure to show that their research is meeting targets set by a funding sponsor. It is remarkable how quickly yesterday's dazzling breakthrough becomes today's humdrum obvious result. The creativity of scientists has been pushed to the limit in the conduct of their research, but also in the realm of grantsmanship.

The point of discussing the pressures on scientists and their potential weaknesses as human beings is most certainly not to vilify them. Scientists did not seek the pressures of the modern university. However, the culture of science leads scientists to downplay the prevalence of ethical issues in research. Rather than showing leadership to help overcome this issue the current system makes universities co-dependents in a system where some scientists succumb to the pressure, end up taking short cuts or deceiving themselves. Moreover, scientific leaders have not confronted the fact that federal ethics regulations are poorly suited to the task of ensuring a high standard of scientific conduct when they put universities in charge of investigating

S. Franzen, *University Responsibility for the Adjudication of Research Misconduct*, https://doi.org/10.1007/978-3-030-68063-3_6

allegations against their own faculty. While blatant fraud may be an example where the "bad apple theory" applies, one cannot ignore that fact that ethical problems can arise through a combination of self-deception and cover up. First, we must understand how mistakes can become something more serious. An honest examination of the origins of bias is required in order help scientists avoid sloppy science. Second, we must recognize that there is little or no education provided to students to deal with this profound issue. Most scientists receive little formal discussion or training in understanding self-deception or possible scenarios that lead to ethical transgressions. Ethical admonitions to avoid the behavior of known bad actors does nothing to help scientists deal with the day-to-day reality of sifting through experimental data to find possible insights while not succumbing to sloppy science. Although observation strongly suggests that the great majority of scientists are honest, there are exceptions in two categories: (1) a very small number who intentionally fabricate experimental results that have no basis in reality and (2) a significantly larger number who delude themselves into thinking that a result means something that it does not. Both can lead to falsification or fabrication, and when that happens both would be adjudicated by the same criteria under the current research misconduct regulation.

6.1 The Pervasiveness of Bias in Science

Despite the importance of the subject, there has been relatively little formal coursework offered in universities on ethical treatment of data. Instead, scientists have relied on an informal process of laboratory experience to teach the idea of verification based on repeatability and reproducibility of experiments to corroborate or discard hypotheses. While individual experiments may fail for myriad reasons, science as a whole is robust, in part because of control experiments and various consistency checks. The power to predict outcomes based on past measurement and experimentation is perhaps the most important consequence of a hypothesis that has become established as part of accepted theory.

Prediction bias is a significant problem in computational chemistry and physics. Prediction bias occurs in complex systems when a scientist tends to look for trends or structure from known systems and predict those in a new situation. For example, predicting protein folding is difficult because there is a handful of common structural motifs, alpha-helix, beta-sheet, beta-turn etc., but there are many other structural motifs that fall outside these standard types. A prediction that relies mainly on the previously established motifs may miss a new type of structure due to prediction bias. In most physical sciences, research is sufficiently complicated that research groups are divided into experimental and computational approaches. It is often difficult to coordinate efforts between these two approaches so that computational or theoretical predictions can be tested systematically by experimentalists. To provide an opportunity for testing of theory there is a biannual competition to predict the structure of a set of proteins from its linear amino acid sequence that have not yet been solved by structural methods. Protein folding has been an important area of research ever since the human genome was sequenced. The Critical Assessment of protein Structure Prediction (CASP) competition is a community-wide effort to compare various methods for prediction of protein folding that began in 1994 and has continued to the present [280]. In this competition an experimental group will agree to withhold their publication for a period to permit other groups to publish predictions of the protein structure from first principles. The concept is that the best way to propel science forward is for scientists to be forced to predict the structure. At the end of the process, the predicted structures are compared to the experimentally determined structure and ranked in order of accuracy.

6.2 Poor Methodology: Cherry Picking, Inadequate Controls and Failure to Replicate

The most extreme cases of scientific bias occur when one selects which experimental results to present based on whether they give the expected result. Choosing the data that agree with one's expectation is known as "cherry picking" and it is a kind of falsification, known technically as expectation bias. However, it is extremely difficult to document that a research group or a scientist has systematically selected data in such a manner. Unlike image manipulation or plagiarism, there is an omission which is not necessarily evident in a laboratory notebook, unless a scientist has tried to change the record after the fact. In some cases, an observer may detect the fact that some data were omitted by using statistical analysis.

The famous Millikan oil drop experiment was designed to measure the charge of the electron. Dr. Robert Millikan won the Nobel Prize for the measurement in 1923. Today this experiment is seen by many as an example of extensive cherry picking, although others defend Millikan's methods. In the experiment, a drop of oil was charged by static electricity while being sprayed into a special vacuum chamber. The deflection of the drop as it passed between two parallel plates of a capacitor was used to determine the charge of the electron. Millikan threw out nearly two thirds of his data because they did not agree with the expected result by a substantial margin. He selected only the data that fell in a certain range. This certainly has the appearance of cherry picking and yet, Millikan's 1913 publication is still regarded by some as one of the great experiments of the atomic age [281]. Nonetheless, during the same period, Dr. Felix Ehrenhaft was attempting to make identical measurement in Vienna using the same type of oil drop apparatus. He did not exclude any data and observed that the results were not reproducible. Ehrenhaft gave up, declaring that experiment did not have sufficient precision to measure the charge on the electron. Millikan persevered, threw out the majority of the data as "outliers" and went on to win the Nobel Prize. Ehrenhaft, who concluded that the experiment could not possibly have the needed precision, was never formally recognized for his contribution [282]. Dr. Daniel Fairbank and Dr. Allan Franklin reanalyzed some of Millikan's rejected data and claimed that the Millikan's method is vindicated by a statistical model [283]. The authors reach this conclusion by comparing the result to predictions from Millikan's time related to the idea of a sub-electronic charge. But this conclusion is a straw man since the sub-electronic charge idea had fallen out of favor by the time of Millikan's experiment. Rather the point of the reanalysis was to show that even though the data thrown out by Millikan were quite noisy, they basically averaged to a value close to an electronic charge. This attempt to exonerate Millikan ignored some of the data as well [7]. The explanation offered by Fairbank and Franklin basically points out that if Millikan had been more honest and included all the data he would have obtained an estimate of the charge of the electron but it would have had significant error. Would Millikan still have won the Nobel Prize for that? Issues such as this one force us to ask what the point of scientific investigation is.

In fact, the cherry picking continued for many years past Millikan's original result. Feynman wrote [284]:

> We have learned a lot from experience about how to handle some of the ways we fool ourselves. One example: Millikan measured the charge on an electron by an experiment with falling oil drops and got an answer which we now know not to be quite right. It's a little bit off because he had the incorrect value for the viscosity of air. It's interesting to look at the history of measurements of the charge of an electron, after Millikan. If you plot them as a function of time, you find that one is a little bit bigger than Millikan's, and the next one's a little bit bigger than that, and the next one's a little bit bigger than that, until finally they settle down to a number which is higher.
>
> Why didn't they discover the new number was higher right away? It's a thing that scientists are ashamed of – this history – because it's apparent that people did things like this: When they got a number that was too high above Millikan's, they thought something must be wrong – and they would look for and find a reason why something might be wrong. When they got a number close to Millikan's value, they did not look so hard. And so they eliminated the numbers that were too far off, and did other things like that

While some scientists feel the need to explain away these tendencies to correct the numbers in accord with a bias, Feynman is matter-of-fact in his recognition that these things happened. Scientists must come to terms with the difference between science as a discipline and the individuals who practice it. As science becomes a worldwide undertaking, we are bound to have ethical transgressions. We must deal with these in an honest and forthright manner to avoid descending into the subjectivity that is plaguing so much social discourse in our society.

It is difficult to generalize concerning the origins of wishful thinking, but the psychology of scientists that leads to bias can be understood from examples. A colleague relayed the following personal story from an early stage of his career. He had been an assistant professor at one of the top universities in the United States. He had discovered an effect that he believed was a new kind of conducting organic polymer. He had submitted a paper to *Science* and it had been peer-reviewed and accepted. As he was carefully reading over the final proofs of the paper prior to publication, a graduate student rushed into his office to show him new data that revealed that the effect was probably due to a contamination of a thin layer of water on the surface of the polymer. Since water can conduct electricity, the contaminating layer of water was sufficient to show the current that had originally been believed to be part of an organic circuit. This would invalidate the entire concept and, worse yet, would be a bush league mistake. The young professor held on to the proofs and stalled for time. He asked the student to conduct a few more control experiments. Within a week it was clear to the professor that the paper was wrong. Yet it had already been accepted in Science! The professor spent a long night thinking about the implications of the accepted paper. He knew that it was unlikely anyone would ever contest the finding even though he knew it was wrong. If anyone ever did contest it, he could probably feign ignorance and it would never be more serious than just a mistake even in that unlikely event. Anyway, by that time his career would be further along, and it would be of less consequence. The next morning, he contacted the editors of *Science* and withdrew the paper. As he did this, he knew that he was most likely ending his chance of tenure at the prestigious university. He would suffer personal humiliation with those colleagues who knew about the submission to the prestigious journal, since he would have to admit his embarrassing error to them. In fact, the most startling aspect of my friend's story was that certain of his colleagues at this prestigious university questioned the wisdom of withdrawing the paper. One might have thought that there would be a kind of understanding or even admiration for the academic integrity shown by the retraction prior to publication. Such did not appear to be the case. With such pressures, one can understand that mistakes are published and then must be somehow maintained or protected if they are not to be exposed [248].

6.3 Abuse of Statistics: Hypothesizing after the Fact, Data Dredging and Ignoring Outliers

The abuse of the statistical methods is rampant in fields ranging from biology to psychology [285]. As discussed in Sect. 3.3, significance testing has been questioned by many scientists with increasing alarm. It is dangerously oversimplified to say that an effect is insignificant if it is less than two standard deviations from the mean of the control group, but then suddenly significant if is greater (perhaps only slight greater) than two standard deviations. Many studies are conducted on very small samples, small numbers of individuals in medical studies or environmental studies. The standard error depends inversely on the square root of the sample size, which is intrinsically limited in many medical, ecological, and biological studies. For example, in a clinical trial it may depend on how many patients have the particular background required and are willing to sign a consent form to sign up for the trial. Many would argue that it is not proper to infer validity for an entire population based on such a test. Despite extensive attempts to provide statistical tests that permit treatment of small data sets, they continue to be an ethical concern.

When results of a study are inconclusive, a sloppy scientist may abandon the original hypothesis and check if there is a correlation that can be ascertained after the fact [86, 250]. The practice known as data dredging or p-hacking is unethical and can lead to meaningless conclusions [286]. If one defines significance as

$p < 0.05$ then, by random chance, one will find one significant correlation for every 20 hypotheses studied. The practice of changing the hypothesis after the fact is dangerous because it gives the appearance of statistically sound result, when the result may be random chance. Studying many variables simultaneously can have the same effect and is known as fishing for a correlation. Such abuses of statistics have led to a spate of failures in epidemiological studies, which can have real-world consequences [285].

Dealing with outliers is one of the great challenges of statistics. Outliers are the data points that are so far outside the expected distribution that they appear to skew it and completely change its significance. Outliers may be mistakes. However, in medicine outliers are often the most important "data points". A virus may kill over 90% of the people who contract it. The survivors, in this case, are the outliers who give clues as to how the human body can withstand the onslaught of the virus. One would be ill-advised to ignore *those* outliers, which we can call *bona fide* outliers. On the other hand, when studying the efficacy of blood pressure medication there may be outliers because of patients who refuse to accurately report their alcohol consumption. If we knew for sure that a patient did not report alcohol consumption, we could exclude the data as a contextual outlier. The inclusion of such outliers may skew the data on the efficacy of a medication for a population of patients. But how can we know the difference between the two types of outlier? The attitude towards outliers and statistical treatment in general differs in various branches of science. There are justifications for exclusion of outliers, but one needs to be systematic and proper about such a practice. Simply throwing away data because it does not agree with the expected result is obviously not justifiable. However, uncritically publishing every single data set when some of the data are flawed may also be a poor practice.

The rise of nanotechnology has increased the risk of selection bias in the presentation of data, both because of the funding involved (public and private) as well as the methods that inherently deal with smaller numbers of observations or statistics on fewer numbers of particles. For example, the widespread use of transmission electron microscopy (TEM) to image nanoparticle structures is an inherently risky approach. If one is looking for an organized collection of nanoparticles, then one must be careful to avoid choosing only the images that contain those collections since these may be outliers in a distribution of structures. For example, I consider an experimental study of the formation of trimers, which are structures of three particles in a triangular structure shown in Fig. 6.1a. These could be formed by a chemical modification to link up the nanoparticles (Fig. 6.1a). To determine if this worked one must consider that there will be a certain number of dimers, trimers, tetramers etc. that form spontaneously, just because of the attraction of one nanoparticle to another. Control experiments to study the incidence of trimers should also compare to a control sample that has no added molecule (i.e. no added red triangle in Fig. 6.1) to determine the number of trimers that form spontaneously in a mixture of nanoparticles as shown in Fig. 6.1b. Since the yield of trimers is always much less than 100% one will obtain a statistical result due to the binding by the molecule (shown as the red triangle) as shown in Fig. 6.1c. But trimers also form spontaneously with

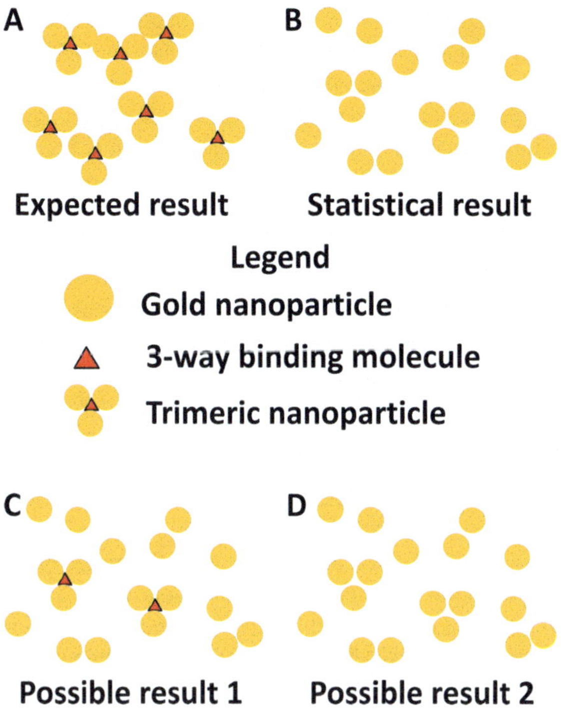

Fig. 6.1 Depiction of possible scenarios that arise from the study of collections of nanoparticles using TEM. The figure illustrates an experiment, in which a researcher tests whether a triangular-shaped molecule can create trimeric nanoparticles. The red triangle represents the molecule and the trimeric nanoparticles that the researcher hopes to see is shown panel (**a**). The statistical result that one would see in a collection of particles with no added molecule is shown in panel (**b**). Particles spontaneously tend to aggregate to form dimers, trimers, and higher-order structures. The reality that the yield of trimers is not likely to be 100% is shown in panel (**c**). The issue of distinguishing spontaneous or statistically formed trimers from those, which are bonded by the triangular-shaped molecule is shown in panel (**c**)

some frequency as depicted in Fig. 6.1d. How does one know whether observation of nanoparticles in the form of trimers is due to the presence of a molecule (symbolized by the red triangle) or simply by chance due to spontaneous association of the particles? The reader should bear in mind that the particles are relatively easy to see using the method, but the molecule that links them together (indicated by the red triangle in Fig. 6.1) is much more difficult to detect and is not observed in the experiment. To establish that there is a statistical difference one would need to compare two sets of particles, one with and one without the red molecule based on a sufficient number of measurements that one could determine the statistical occurrence of the trimers for comparison with the chemically bound trimers. Thus, one is left with the question, how would we know that the binding molecule is present and that the trimers are formed because of the desired reaction?

While the need for control experiments may seem obvious, it is most essential to train students to detect and avoid bias. Assuming for a moment that there is no intent to deceive or misrepresent data, we can understand how collective bias is introduced as follows. If a hypothesis has been articulated that a chemical compound can cause nanoparticles to form structures, such as the trimers shown in Fig. 6.1, the inexperienced student sent to conduct TEM experiments might interpret the instructions as a mandate to find the expected structure. Then the student selects the regions on a TEM grid where more of the structures are found to show that the experiment was successful. It is important to understand that the region of a TEM that can be studied in a single experiment is on the order of one hundred square micrometers (e.g. a square ten micrometers on a side). If the grid has an area of approximately 1 square centimeter, then the area selected corresponds to one 1,000,000th of the total possible area for study. The student may ignore other regions of the grid which are less interesting but may be more representative. The student may reason that the research advisor will not be interested in the other regions on the grid that do not have the intended structure. If the advisor does not ask specifically how the data were selected, then the advisor may believe that the structures are representative of the majority of grid, when this may not be the case as all. There is a danger that this can turn into a kind of academic "don't ask, don't tell" policy. The advisor does not ask for details and the student does not tell exactly how the measurement was made. If a student selects trimers and takes TEM images of them, while ignoring other structures, cherry picking has occurred. The damage to science is already done and there is little more to say. The data will be reported incorrectly, as though trimers form in high yield, but this is a false picture. Current regulation holds that research misconduct occurs only when such an action is conducted knowingly or recklessly. If the graduate student is following the directions of the advisor, believing that there is a justification for the selection then it is the advisor who should be held to account. However, the advisor often does not see the raw data and only knows what the student presents in a group meeting. Thus, the advisor may also not be consciously selected the falsified distribution. According to the current definitions, this type of misrepresentation that involves a misunderstanding is not intentional. Scientists should ask the tough questions and misunderstandings between student and advisor that lead to a false conclusion is obviously to be avoided. Although it is a self-deception, it is not research misconduct. This example gives the reader a sense of the problem today. The hierarchy of scientific research shields many professors from knowing how the sausage is made. The professor explains the goal and sends the student to find data that correspond to that goal. The professor should ask critical questions when the student returns. However, data that conform to expectations are often so exciting that the conversation is about the next experiment or how to write up this result for publication.

When continued over time such a procedure risks further consequences. If a structure is perceived to have commercial value, then the activity just described may lead to an invention disclosure or patent. Funds may be invested. At that point, the entire project takes on a new dimension and the sloppy science may become fraud. Should it be considered fraud by accident or by intent? In the example in Fig. 6.1, we may imagine that the trimer is thought to have value as a component in a circuit used for molecular electronics. If the researcher claims to have a procedure for creating trimeric particles in a reproducible fashion, a reader or reviewer may picture the result in Fig. 6.1a. If the scientist never mentions the possibility of a statistical distribution, we may call that sloppy science. On the other hand, if the scientist claims that the result is overwhelmingly the trimeric nanoparticles shown in Fig. 6.1a, when in reality there is a distribution (Fig. 6.1c) or possibly no formation of trimeric nanoparticles at all above the statistical background Fig. 6.1d, then the claims are misleading at best. There have been hundreds of papers published that claim that certain nanoparticle structures have been made. Sometimes these studies are careful to document all the possibilities and report the control experiments, but there are numerous examples in the research literature where control experiments were not mentioned. It was assumed that the careful work was done, but the exciting result drew the headline. Ironically, for many years, this type of publication was more common in highly prestigious journals.

Another concern is that many studies are not replicated and yet conclusions will be made that are proposed to have general validity. One must be concerned about any result of study has been conducted only once even if the p-value is low enough to imply a statistical significance [287]. Despite the issues of sample size and lack of confirmation, authors will almost always present a conclusion from the study to justify publication. Dr. John Ioannidis expressed concern for the validity of such studies in a provocative, highly-cited article entitled "Why most published research findings are false" [287]. Ioannidis wrote, "Several methodologists have pointed out that the high rate of non-replication (lack of confirmation) of research discoveries is a consequence of the convenient, yet ill-founded strategy of claiming conclusive research findings solely on the basis of a single study assessed by formal statistical significance, typically for a p-value less than 0.05." Simulations based on the considerations above provide potential explanations for the lack of reproducibility of research results. The point is that regardless of the sample size, the result of a single study conducted by one research group may succumb to bias. Replication by other research groups is the ideal in science. Failure to replicate complex studies in medical, biological or environmental science is the norm since such studies are expensive and time-consuming. In fact, the funding agencies do not require replication and the competitive funding climate makes it unlikely that there will be a change in this way of conducting science anytime soon.

Expertise in statistics is not required for a PhD in many scientific disciplines. The rush to the laboratory in graduate school in certain disciplines does not foster a broad understanding of the issues in the treatment of data. The deficiencies in statistical education is one example that is indicative of a more general issue that needs to be confronted if we are to honestly deal the current trends in scientific reporting. The most serious falsifications may never be detected because they are so embedded in the procedures, which are hidden from view in laboratory notebooks that no one will ever see them. Some fraction of unethical publications may arise out of ignorance, but clearly such a state of affairs enables an unscrupulous person to commit fraud. If they are caught in their fraud the person can feign ignorance in some cases because of the complexity or interdisciplinary nature of the research. Improper statistical reporting is one avenue down the slippery slope from carelessness to self-deception and ultimately to intentional fraud. How common is this type of behavior? This is a crucial question we must answer to understand the extensive lack of reproducibility and spate of retractions that have been documented in science in recent years.

6.4 When Does Self-Delusion Become Intent to Deceive?

Most scientists find it difficult to imagine that anyone would persist in a claim once it became obvious that the evidence does not support the claim. Yet, there are documented cases of controversies where self-delusion leads to an observation that the researcher defends even in the face of overwhelming evidence to the contrary coming from numerous research groups. The famous cases of this type are worth considering because they exemplify how reputable scientists can refuse to see a self-delusion [288]. In the late 1960s, a new form of water, so-called polywater, was reported by scientists in the Soviet Union. Dr. Derjaguin was an established scientist when he began to pursue the idea that polywater existed. The theory of a new kind of water structure, known as polywater was based on the observations that water that collected in narrow tubes (capillary tubes) froze at much lower temperature and boiled at a much higher temperature than normal water. Experiments carried out subsequently showed that polywater contained impurities. Thus, the effects of freezing point depression and boiling point elevation are textbook examples of the equilibrium shift of phase transitions when salts are dissolved in water. Still, the proponents of the theory continued to believe that they had discovered a new phenomenon for a decade or more [289]. A second example of a highly publicized scientific self-delusion came to light in a 1988 publication in *Nature* by Dr. Benveniste claiming that the activity of an antibody could be observed after dilution to the point that there was statistically less than one antibody molecule per container. This led to a controversy over the observed effect of biological molecules even after they had been diluted so many times that it was statistically improbable that even a single molecule would be in the sample [290]. The explanation was said to be in the "memory" of the presence of the molecule (here an antibody) by water. When the editor of *Nature*, John Maddox traveled with two observers to look at the procedures using double blind experiments, they found that the effect vanished. Dr. Benveniste never accepted this conclusion. This incident has many parallels with the case of Blondlot's N-rays, which took place over 80 years prior. In both cases, a curious editor traveled to the lab of a French scientist and spent enough time to obtain evidence of a self-deception. When the evidence was published neither of the scientists who was exposed could admit that there was a self-deception [291].

These examples appear to conform to the notion that mistakes in science will be challenged by conflicting accounts that will ultimately clean up the scientific literature [87]. But in reality few scientists have the means to repeat complex published experiments or travel budgets that permit them to visit each other's laboratories just to check their work. Even when scientists attempt to correct the record it is all but impossible to convince someone that they have engaged in self-delusion. It may be a question of national pride (e.g. Blondlot's N-rays) [291], political motivation (e.g. Lysenkoism) [129, 292], financial gain (e.g. Wakefield's correlation between vaccines and autism) [293] or personal pride (e.g. Derjaguin's polywater) that prevents a person from admitting that their theory has no firm basis.

There are many examples of elaborate deceptions such as the Aberhalden theory of defensive enzymes, which was a serious self-delusion that affected an entire generation of scientists. The Swiss Biochemist, Prof. Emil Aberhalden proposed that enzymes were stimulated by the immune system, was based on a pregnancy test that turned out to be completely fraudulent. Although the falsification was debunked at an early stage [294], it took 93 years for the fraud to be finally recognized [295]. A surprising case involved the American Nobel laureate Alexis Carrel, who won the Nobel Prize for developing the technology that permits removal of cells from a tissue in an organism and to a broth where they can be kept alive and studied. This part of Carrel's work is a great contribution to science. However, Carrel had notions of how to maintain cells in such culture for an indefinite period. On January 17, 1912 his research group transferred some embryonic chicken heart cells to cell culture and observed in cell culture for 34 years, which is five times longer than the age of a typical chicken [296]. However, there was a significant flaw in the experimental protocol. The cells needed sustenance and the broth in the cell culture was periodically replenished with fresh broth so that chicken heart cells were periodically introduced into the culture each time the new broth was added. It did not occur to Carrel or his co-workers that they were not carrying out any control experiments. This is such a high profile and yet careless mistake that it is considered a fraud by essentially all modern observers.

Indeed, if we look at the totality of science from its early roots, we can see a strong tendency for manipulation of data in even the most venerated of scientists. In an article entitled "The Golden Age of Fraud" Dr. Walter Gräzter describes numerous cases of fraud starting with Sir Issac Newton. In Newton's case the issue centers on the numerical accuracy of his theories [296]. Dr. Robert Westfall carefully described how Newton's reported accuracy of both the acceleration of gravity and the speed of sound exceeded any reasonable ability of the theory to fully account for the effects [297]. Regardless of the motive, Dr. Charles Babbage, who also held Newton's chair at Cambridge in the early nineteenth century, pointed out the exaggerated claims of accuracy in Newton's mechanics in the third volume of the Principia. Babbage's treatise "Reflections on the Decline of Science in England" published in 1830 discusses practices such as data trimming (cherry-picking) and forging (fabrication), which are still part of the fraudster's arsenal today [298].

In examining the motive of the scientist who uses aberrant practices, such as fudge factors, lack of control experiments, failure to use single- or double-blind methods etc., when they should know better, we can ask to what extent the scientist's awareness has gone beyond self-deception to a conscious awareness that they have failed to conduct either experiments or analysis in a proper manner. Many of the historical examples of pathological science can be ascribed to self-deception. Scientists are supposed to constantly let experiments be the guide. Regardless of this expectation, a great deal of science is conducted to confirm theories that are *assumed* to be correct. In such cases, some scientists may be willing to ignore data to obtain agreement with the theory. Many mistakes are made by seeing results that one wants to see and disregarding those that do not agree with the theory.

For centuries, the motive for unethical practices in science was to give the appearance of having completely solved a problem in order receive credit for being first to develop a self-consistent theory. Loose ends or messy data with many outliers could give the impression that the theory was not explained. Although more serious misrepresentations have also been conducted by scientists such as the misrepresentation of methods by Pasteur [7]. Pasteur's germ theory was a great success, but in the effort to keep ahead of the competition there was one set of experiments where he used a competitor's method, but claimed to have used his own method. Pasteur's scientific achievements have stood the test of time. His methods worked and his ethical lapse was mainly an expediency since he had little time to produce a vaccine and his own method was still at an early stage of investigation. It was the pressure of the competition that compelled him to hide the fact that he had used his competitor's method in that one instance. Since Pasteur's competitor was German, there was also concern for national pride and perhaps even national security that drove Pasteur to commit a deceptive act. While the historical problem of falsifying to maintain the image of being a pioneer in the conduct of science no doubt still exists, the issue is currently exacerbated by a new set of issues in the consciousness of scientists: funding. Since funding is essential to the very survival of a scientist the new motive to find agreement between the theory and experiment is even more compelling. The specialization and complexity of modern science have increased the possibilities for obfuscation to give the appearance that a study is properly conducted when in reality it is flawed.

The insidious ways that corporate funding and the startup company culture have influenced universities have been documented by Jennifer Washburn in the book "University, Inc." [4]. Scientists who start their own companies often delude themselves into believing that they can run an academic research group on the same technology without any bias. Companies can communicate the desired result of studies to their employees. Admitting that one has failed to obtain a conclusive result or that the result is negative can result in investor flight. The first act of trying to present a result as more positive than it is or deciding to omit a figure because it "does not look good" can lead to other steps, all justified by the need to support the research. Similar pressures are common in any academic research where grants are up for renewal.

Few scientists can ignore the pressure to write grant proposals in such a way as to attract attention of the program officer and get one's research to be at the top of the grant pile at the funding agency. If a scientist begins to write creatively about what has been done in the laboratory in an attempt to justify funding, there may be liberal interpretations of the meaning of success or validation of hypotheses used to impress reviewers. Grant proposals are not always peer reviewed for accuracy, but are usually reviewed for the promise of an idea. Feasibility is also considered, but research scientists are given wide latitude because funding agencies do not want to impede innovation. Reviewers are asked to evaluate whether the research is interesting, topical, and feasible based on the preliminary data presented. Some research scientists understand this process and present preliminary data in such a way as to appear more conclusive than they really are. The line between what has been done and what one would like to have done is sometimes blurred in grant proposals in ways that would not occur in most peer-reviewed journal articles. Marketing has become accepted as a necessary evil in a world that used to be reserved exclusively for hypotheses and data. Federal regulation does not address inaccuracies such as exaggeration unless they lead to clear falsification or fabrication of the research record. One danger of marketing is that the PI who has made projections about what can be accomplished will feel pressure to validate the hypothesis advanced with anticipation in a funding proposal. The path to misrepresented science is not necessarily based on obvious distortions of data but may consist of subtle issues of experimental design or interpretation. Over time the issues lead to discrepancies, which scientists may not want to admit.

Even public utilities and federal agencies have been guilty of misrepresenting science to defend their interests. Much of the effort to monitor food, air, water, soil and contaminants in general is not science in the sense of hypothesis-driven research. Instead, these are analytical studies that are subject to interpretation, can be the subject of contention among stakeholders. They can and often do involve university research scientists who validate or confirm methods and interpretations. As an example of how contentious the presentation of data can become arose in a study published by the Centers for Disease Control (CDC), claiming that there was no evidence that consuming elevated lead in water of Washington D.C. 2001–2003 elevated blood lead of residents above levels of concern, the CDC research was used to downplay health risks of lead in water exposure. Over a period of 5 years, Dr. Marc Edwards, Professor of Civil Engineering at Virginia Tech found elevated levels of lead in children in Washington D.C. During Edwards investigation the CDC issued press releases attempting to discredit Edwards, refused to fulfill Freedom of Information Act requests. Edwards investigation also revealed flawed or fraudulent reporting in publications authored by the US EPA, a George Washington University professor working for the local water company and the water industry. The CDC report was ultimately discredited by a bipartisan investigation, which called the CDC's effort "scientifically indefensible" [299]. Edwards argues that the failure to learn from the D.C. tragedy, made future problems with lead in water inevitable elsewhere, including ultimately the 2014–2015 Flint MI water disaster which Edwards also exposed [300]. Although the perpetrators of the D.C. tragedy and the authors of the falsified scientific and technical reports were never held accountable, several whistleblowers who attempted to alert their superiors and the public to the situation were fired. Such cases illustrate that great harm can arise from fraudulent, falsified or shoddy science. Regardless of the origin of mistakes (i.e. negligence or intentional) these examples show how each bureaucracy involved had engaged in a cover up, even at the expense of public health, not to mention scientific accuracy.

6.5 Retractions as a Symptom of the Problem

Confidentiality regulations make it impossible to obtain quantitative data on the number and outcome of research misconduct investigations. Instead, studies of the retraction rate have been used to present a pervasive and increasingly serious problem [301, 302]. ThompsonReuters data show that retractions were up by a factor of 15 in the decade starting in 2001, while the number of publications increased by 44% [303]. Retraction Watch was founded in August 2010 to cover retractions, corrections, and other aspects of scientific and academic publishing. The stated goal of the website is to promote greater transparency. The rapid growth of Retraction Watch has surprised many observers and even the founders, Ivan Oransky and Adam Marcus. There have been more than 2000 posts since 2010 and essentially as many retractions. The readership of Retraction Watch has grown to more than 150,000 unique visitors each month, generating more than 600,000 page views. The website has also provided a forum for the expression of opinion. There are comments on retracted papers as well as details about problems in papers where ethical issues are suspected. In some cases, the comments of readers have provided a basis for wider investigations into the conduct of researchers. Most of the comments show that readers are concerned about the ethical issues and thus the website appears to have touched a nerve in the academic community [304, 305].

Not every retraction is an indication of an ethical problem with an article. The authors of Retraction Watch have made a point of highlighting the numerous honorable retractions. However, a recent study of 2047 retracted articles in the PubMed database revealed that 67% were attributable to misconduct [1, 306]. Although only a small fraction of the articles containing falsified or fabricated data are made public, a recent study attributes 27% of all retractions in the scientific literature to falsification or fabrication [307]. The discrepancy between these two studies is most likely due to plagiarism, which may comprise as much as 40% of the retractions. The retraction rate is increasing in part due to the increased detection of plagiarism using modern search engines [1]. The number of serious cases of scientific fraud is increasing [308]. Given the policy of confidentiality and the current climate of damage control in university administrations, the cases that we know of are the tip of the iceberg. For the cases we know of a recent study of 266 cases found that 14% of cases the charges are dismissed for lack of evidence while some kind of action was taken in about half of the cases. Without oversight or established procedures for how evidence is to be treated one must ask whether dismissal of cases is simply because researchers are given the benefit of the doubt and no one actually checks the complex details [309]. Since real fraud can be quite sophisticated the complexities of corroboration of results often must be checked to find it [310]. Surveys find a significant underestimation by university academics of the incidence and seriousness of the problems of falsification and misconduct [311]. The available evidence suggests that both retractions and incidences of falsification/fabrication are increasing. This is clearly a concern, but it is the part of the problem that is hidden from view by confidentiality that is the most insidious.

6.6 The Rise of Search Engines and the Fall of Plagiarism

Since evaluation of faculty for promotion and funding relies on the number of papers whose content will not be evaluated since colleagues are too busy to read them, there is an incentive to pad one's curriculum vitae with papers that contain essentially the same results as previous work. Recycling results and text is known as self-plagiarism or dual publication [312]. While plagiarism used to be one of the most common forms of scientific misconduct, the advent of new technology to detect plagiarism and dual publication has begun to reduce its incidence. Search programs such as ETBlast can identify similar text in millions of articles in a few seconds [313]. ETBlast is based on the Blast (Basic Linear Alignment Search Tool) algorithm, which was originally used for searching genetic and protein sequence databases for nearly identical sequences. The idea behind this code is to find strings of letters that are not necessarily 100% identical but may have a few changes or gaps. There is a score for similarity of sequences, or text in the case of ETBlast, that is based on how many characters are identical and how many gaps exist in the text. The authors of ETBlast have further developed a website called Déjà vu that exposed the dual publications of several prominent authors [314]. While this public service appeared to be off to an excellent start it suddenly was closed down. The approach is still available commercially with the name HelioBlast.

Detection of plagiarism has become a business, which raises the question of whether there is now a conflict of interest in the zealous enforcement of plagiarism at universities. Using software packages such as CrossCheck, Turnitin, PlagiarismDetect, WriteCheck and so on, recycling of text from older papers is much more likely to be caught [315]. So too is the careless college freshman or recently arrived graduate student from China who are not aware of the precision with which the plagiarism rules are enforced today. The ease of detection has turned plagiarism into a reporting and enforcement issue. Nonetheless, plagiarism whistle-blowers can suffer retaliation like any other [304]. While it is a positive development that plagiarism can be detected and may even be on the decline, this turn of events does not mean that falsification and fabrication have abated.

The zeal with which plagiarism is punished in universities betrays a bad conscience with regard to transgressions that are more difficult to prove. Incoming freshman in universities are admonished not to write more than ten consecutive words from a source without proper attribution. Failure to comply constitutes grounds for expulsion. Certainly, we want to encourage responsible citation of references, but this type of dire warning aimed at incorrect citations misses the larger issue. One can plagiarize ideas without necessarily using the words verbatim and sometimes one can use words verbatim that have no intellectual novelty or artistic value. For example, to designate the reused descriptions of experimental procedures as "plagiarism" distracts from the important issues of theft of ideas or innovation. In many kinds of research, the methods section discusses standard techniques and contains few, if any innovative ideas. The intent of the plagiarism regulation is to prevent idea theft or recycling of publications. The search for every possible phrase with more than ten consecutive words has become so extreme that some international students can find themselves facing expulsion for minor infractions due to a lack of familiarity with American standards of referencing. In cases where expulsion is threatened there is an implicit assumption of intent to deceive that actually should require an investigation. If the alleged plagiarism occurred in federally funded research an investigation would be required. While a faculty member's rights will be respected the foreign graduate student or post-doc may face a different fate. None of this is to deny that major plagiarism and recycling of papers is a problem, but rather to point out that the technology has become so powerful that we need to be careful not to make up for deficiencies in other areas by excessively narrow definition of plagiarism that sweeps up students who make small mistakes in the same net as those who commit wholesale theft of others ideas and text.

The plagiarism of grant proposals and research ideas within a research group is still a potential problem since the documents are usually not in any database and this type of misrepresentation is not normally detected by the searches using software or on-line technology. The databases available for search typically include journal publications, patents, and books, but submitted (and unfunded) proposals are confidential. The exception to this is that agencies can scan submitted grant proposals for plagiarism within their own databases and check to see that peer reviewers have not plagiarized from grant proposals they reviewed or that PIs are not submitting grant proposals that are recycled from their previous submissions. The more general issue of theft of other's research ideas is still difficult to detect because funding agencies do not yet crosscheck the databases of other agencies or private foundations. Peer reviewers have the opportunity to read proposals prior to the funding decision. Research that is not funded is supposed to remain confidential. Mentors often see their former post-docs research proposals for the purpose of helping them improve their writing. These documents are also to remain confidential. The violation of trust and theft of ideas in either category is a serious infringement of the rights of others. This form of plagiarism still relies on detection by the injured party.

6.7 What Lies Behind the Reproducibility Crisis?

The Retraction Watch website has contributed to an awareness among scientists about the lack of reproducibility in scientific publishing [316]. Just like retractions themselves, lack of reproducibility calls into question whether studies were done properly and even, in some cases, whether they were done honestly. According to the ideal of the scientific method scientists will carry out their experiments with sufficient care that others can reproduce them, but there is a great deal of evidence that the reproducibility of university research is quite low [317]. Lack of reproducibility does not necessarily mean that the research conclusions are incorrect. It could simply mean that the description of how the work was conducted is incomplete in some way. That type of omission is not always innocent. There may be intent to withhold certain information to maintain a competitive advantage in the funding race. Regardless of the reason, the lack of reproducibility represents a major problem that challenges the premise in science that we can build on previous experience.

The issue of whether the incidence of irreproducible research is increasing or decreasing, whether it is a major or a minor problem is not settled. While one recent study downplays the concerns [318], there have been sufficient reports of serious loss of confidence in the standards of academic reporting that one should take it seriously. A report from Bayer Corporation revealed that there is a pervasive problem when industry studies failed to reproduce 67% of university studies [319]. In an article in the Wall Street Journal entitled "Mistakes in Scientific Studies Surge" [303], journalist Gautam Naik documented serious errors in studies that have ramifications for human health. A study in the journal *Lancet* had provided evidence that a combination of two blood pressure medications was more effective than either alone. Over 100,000 patients were given the new prescription. Then after a period of 7 years, the study was retracted because of apparent data manipulation [320]. The combination, it turned out, was not an improvement, but it had put patients at greater risk for life-threatening side effects. However, not only is the damage done in terms of the patients already affected, but it is also difficult to change the habits of doctors after years of a particular practice. Of course, this is a case where the paper was actually retracted, which does not always occur when serious flaws are exposed. For example, Amgen spent millions of dollars on studies to reproduce the anti-cancer effects of the protein STK33 and failed to reproduce the basic findings reported in the published literature [321]. Research scientists at Amgen then studied academic research on cancer therapy and concluded that 89% of the academic studies could not be reproduced! [322] In an article entitled "In cancer science, many "discoveries" do not hold up", Sharon Begley discusses the surprise among the Amgen scientists as they made this finding. The 53 papers in the Amgen study were not chosen at random, but rather were considered "landmark" studies from top academic laboratories. Perhaps this is not so surprising when one considers the complexity of any study of living organisms and the number of factors that may affect the outcome. At a minimum, these findings are sobering because they call into question the assumptions of reproducibility that are taught to each new class of science and technology students. If the origins of the lack of reproducibility are really complexity, then perhaps we need a more realistic assessment of way science really works. This is part of the message of the recent study entitled "Is science really facing a reproducibility crisis?" [318].

"The Reproducibility Project: Psychology" was the first large scale project to test the reproducibility of academic research led by Dr. Brian Nosek at the University of Virginia [323]. One method for estimating reproducibility was to check whether the p-value of the replicated study was less than 0.05 ($p < 0.05$) and shifted in the same direction as the original study. Based on this criterion only 35 of 100 studies selected were found to be reproducible [324]. The authors point out that this is somewhat arbitrary and that some studies were remarkably close to 0.05. Using a complementary method to test whether the replicated study was within the 95% confidence limit of the original study 47.4% of the studies were replicated. The Reproducibility Project: Cancer Biology is a systematic study of reproducibility by a large team of research groups that studied the reproducibility of 50 publications in cancer biology, which is a follow-up to the report by Amgen [317]. Reducing a complicated study of this type to a single number is always fraught with peril. In addition to comparing significant (p-value) one can also use the statistical power, which is the probability that an effect will be detected when it is actually there. A recent study of 246 neuroscience publications found a statistical power of studies to be 21% [325], which means a true effect will only be estimated to be true for 21 of every 100 studies. The main reason found for the lower statistical power is that the sample sizes were small. Moreover, the authors noted that sample bias is a persistent concern, particularly when sample sizes are small. By any measure used, the reproducibility of published research is sufficiently low to cause great concern for anyone trying to evaluate the state of knowledge in biomedicine, psychology or a range of other disciplines. The origins of this lower level of reproducibility may be poor reporting, incomplete descriptions of methods or other errors that are fraught with ethical issues. Sometimes a sample size is small because a research scientist is inherently limited, but if there is an adequate population and yet the researcher chooses to study a small sample, this may be due to laziness, expense or bias. There is a growing number of scientists who have acted via the internet to question poorly documented, statistically flawed or obviously manipulated publications.

A new statistical tool developed by Dr. John Carlisle for reanalyzing published studies was recently applied to 5087 clinical trials conducted from 2000 to 2015. One conclusion of the study was that 1% of the studies contained errors outside statistical norms [326], *which indicates these studies are fraudulent (according to the definitions of the study).* Dr. Margaret MacDougall approaches systematic statistical analysis for the detection of fraud in biomedical research [327]. It is up to the journal editors and reviewers to catch certain obvious misrepresentations. For example, study 329 of paroxetine [328] was negative for all eight clinical outcomes and positive for harm. Yet, the conclusion of the study misrepresented the science and concluded that the drug was "well-tolerated and effective" [329]. The statistics were in good order, but the conclusion of the study had no relationship to the statistical conclusion. Statistics can appear to be sound in studies based on data dredging, but the correlations are meaningless. Such studies have a significant negative impact on epidemiological studies [250, 285]. Many observers are suggesting that peer review, standards of scientific publication and the terms provided for clinical trials all are in need of reform [330]. There is a sense of urgency in the field of medical science because the conclusions of studies have impact on the therapies for treatment.

The fact that science has grown in complexity is perhaps to be expected. However, the point is that the issue of reproducibility provides cover for those who are sloppy or consciously trying to falsify scientific data. The fact that a scientist can claim that conditions are different in a different lab can be used to defend against any experiment no matter how simple that exposes a fraud. Lack of reproducibility is troubling in itself, but it is particularly disconcerting that it permits a new narrative of "alternative facts" even in the world of science. Clearly, scientists need to think about how to measure reproducibility and confront the genuine lack of reproducibility when it is observed, but this issue increases the urgency of finding a means of adjudication of allegations of falsification that is swift and fair. We need to separate those publications that are either sloppy or intentional frauds from the work that may be sound, but suffer from omissions, lack of clarity or differing conditions.

6.8 Self-Policing of Science on the Internet

One solution to issues of integrity in science would be for scientists to police themselves. Unfortunately, scientists as a group have not taken on this difficult task and there is not a generally accepted mechanism for this type of oversight [331, 332]. Although they lack formal approval by the scientific community, several websites have recently provided avenues for participants to question the validity of published results. The site Science Fraud existed for 6 months and provided a forum for public whistleblowing. Legal threats followed when the identity of the author of the site, Paul Brookes, was exposed [333] and the website was immediately shut down [334]. In an interview Brookes explained how he became frustrated with the lack of a mechanism for correcting obviously erroneous results in the scientific literature. Apparently, he was not alone as nearly 300 articles received significant attention within the short time span that the website was operating.

PubPeer was founded in 2012 and has become influential in the debate over retractions [267]. This site is the first in a new trend of post-publication peer review (PPPR) websites [335]. Numerous retractions have resulted from information that initially was posted on the website. PubPeer permits anonymous reviewers to point out inconsistencies and flaws in data, figures or interpretations of published work. The founder Brandon Stell decided in 2015 that he could no longer remain anonymous, although the website has gone to court to protect the anonymity of informants [336]. The site was recently sued by a medical faculty member, Dr. Fazlul Sarkar from Wayne State University who alleged that he lost a job offer after his work was criticized on the website [337]. Sarkar sued PubPeer to attempt to force publication of the name of the informant. However, Sarkar lost the lawsuit, which further exposed evidence of his research misconduct conviction at Wayne State University [338]. At the present time 19 of his publications have been retracted. At present, there are more than 35,000 comments on PubPeer. The high level of activity on these sites and the corresponding legal challenges illustrate the charged atmosphere that exists today as scientists push an increasing number of publications into the scientific literature that contain questionable findings and other scientists feel the need to expose the fallacies in those publications. Comments on PubPeer suggest that scientists are acutely aware of the institutional failure to adequately deal with the problem of falsification. Recently, some journals have permitted post-publication comments on journal websites. These are subject to the moderation of the editors and may therefore be subject to bias, but it appears that the effect of PPPR has been felt and journals are beginning to respond. Some scientists are sufficiently concerned about the current ethical situation that they have become post-publication vigilantes who actively search for flawed studies by looking through journals [339, 340]. It is undeniable that at present there is an epidemic of retractions in the biomedical field and many of these are associated with various kinds of misrepresentation ranging from image manipulation to outright doctoring of the results. Many of these cases raise serious questions about the mentoring students are receiving or even suggest that the PIs themselves are complicit in falsification.

A few examples serve to illustrate that this problem is both widespread and high impact. In 2013, articles in *Nature* and *PNAS* from the group of distinguished professor Dr. Maya Saleh of McGill University were found to have falsified images. On the Retraction Watch website Oransky wrote "The committee said they could not determine who had falsified the figures, but said that there was no evidence it was Saleh, who was the only common author among the four papers." [341] In an article titled "Former UT-Southwestern cancer researchers faked data in ten papers: ORI "Oransky describes how post-docs in Dr. Adi Gazdars' laboratory at UT Soutwestern falsified data in two publications before they left for research positions in Japan [342, 343]. However, the lab has multiple issues on Retraction Watch some of which were detected by readers. With regard to the nine retractions by Dr. Cori Toth of the University of Calgary [344], Oransky wrote "Outside observers were, however, more thorough than the faculty in assessing the problems with Toth's studies. Subsequently, Retraction Watch identified other questionable data' and a complaint triggered a university 'Committee of Investigation' process" [345].

6.9 The Role of Research Group Culture and the Responsibility for Misrepresentation

The scientific community has difficult time being objective about falsification or fabrication committed by senior research scientists. Cases involving junior scientists essentially always result in expulsion of the respondent to allegations. In certain famous cases, the fraudster was personable, talented and sought out a famous mentor (e.g. Breunig or Darsee). These publicized cases may be important because of what they tell us about perceptions by the community. The scenario for misrepresentation by the junior scientist often involved a mentor with many responsibilities who did not have time to supervise their protégée or was separated by distance so that it was not possible to observe the fraudulent practices. But the extremely high productivity should have made mentors suspicious. Robert Sprague and professor and whistleblower in the Breunig case was a rare mentor who detected fraud in a long distance collaboration. Dr. Sprague blew the whistle and paid a high price for that decision [346]. A description of the perils of being a whistleblower and the aftermath can be found in an article by Dr. Sprague entitled "Whistleblowing: A very unpleasant avocation" [346]. Sprague discovered that Stephen Breunig was fabricating data in clinical studies, which was so serious that it resulted in patient abuse. When Sprague reported this to the HHS he found that that he became the suspect and suffered a loss of funding. For years, no one took the allegation seriously. While it took the HHS an unacceptably long time to realize their mistake, Breunig was eventually prosecuted and went to prison. Although Sprague was vindicated, it was a great cost to his scientific career.

The culture of a research group is a very important aspect of the training that students receive. The current system of patronage in science gives essentially all the credit for any discoveries to the mentor. There is a great hypocrisy in the current academic system because much of the misrepresentation is blamed on junior scientists. It is up to mentors to ask hard questions that would expose inconsistencies in research. While one can imagine a scenario in which a student actively tries to deceive a mentor, a vigilant and inquisitive group culture should expose errors. A research group should be discussing science and there should be active challenges to data presented by others. There are potentially serious problems with the culture in the research group in publicized cases of serial falsification.

For every major case of fraud there are numerous cases of falsification or image manipulation. These require retraction, but it is likely that these cases are complicated by the possibility of poor education or background factors that could have been corrected. In many of these recent cases graduate students have been held to account despite nagging questions about the role played by the advisor. For example, six publications were retracted from the Emory University research group of Dr. Alexander [347]. Former post-doc Lian Zuo admitted to multiple image manipulations in three of the articles. However, recently an article [348] was retracted from the same group and Lian Zuo was not a co-author [349]. Although the Emory University administration refuses to comment, the record shows that six articles [348, 350–354] have been retracted for image manipulation. How much of this misreporting is poor training? How much is misunderstanding? How do we assign responsibility in these cases of group research? Since nearly all research is done in groups, these are burning questions for young people choosing a career in science. Despite an overwhelming number of falsifications reported on Retraction Watch, that website is documents only a small part of the current problem. Although PIs receive the credit for any breakthrough, PIs most often claim that they had no knowledge of misrepresentations discovered by others in their research group or even by readers of their publications. Regardless of the view of scientists on who bears the responsibility, the university administration has a vested interest in protecting faculty. The students and post-docs on the other hand are usually left to fend for themselves.

The issue of responsibility for falsification committed by group members of a research group is debated on the Retraction Watch webpages and on many science blogs. In well-documented cases, graduate students or post-docs are alleged to have falsified data in publications that also name the professor and PI as an author (usually the corresponding author). While many observers who write comments on Retraction Watch have no sympathy with the PIs, there is a minority who point out the pressures on the PI. Indeed, the job of professor is a challenging one with many responsibilities. However, this defense must take into account that the PIs have chosen themselves to be ambitious enough to run a large laboratory, often so large that they do not have much personal contact with everyone in the laboratory. In the case of Dr. Alexander from Emory University there were six retractions with more than one senior author who had committed the image manipulation [347]. It seems clear that the PI bears at least some responsibility in such a case. The PI is responsible for the culture in the lab and for checking the details of the publication.

A university can decide to punish the graduate student and exonerate the PI. If a university does not wish to explain the reasons for their decision, a university *unilaterally* determine its jurisdiction in a research misconduct case [111]. In a research misconduct investigation at the University of Utah, graduate student Rajasekhar Anumolu in the research group of Leonard Pease was investigated for allegedly altering two figures, which led to the retraction of two research publications [355, 356]. Since the published manuscripts cited funding from both NSF and NIH, either the NSF OIG or the NIH ORI would be required make the final determination and approve the investigation report from the university. However, the university notified the two agencies that neither of them had jurisdiction over the case. The university stated that it had conducted a preliminary inquiry and established that none of the funding had come from federal sources despite the fact that the journal articles acknowledged federal support. Apparently, the agencies were content to permit the university to decide that they had no jurisdiction even though both NSF and NIH funding were cited in the acknowledgments of the manuscripts under investigation! In theory, the agencies could have contested this, but in practice the regulation gives the university the primary responsibility and the agencies essentially always defer to the university decision. If the university states that no federal funding was used, there is little incentive for an OIG to question that statement. Once it had established sole jurisdiction, the University of Utah determined that the student was guilty of research misconduct and that the professor was exonerated. First, the university is protecting its investment in a faculty member with an active research program that brings in overhead. Second, the university would like to ensure that it will not be required to return any federal funding. However, there is little more anyone can determine about this case since there is no information available to the public other than the fact that the foreign graduate student was expelled from the university. This is a case where one can read obvious frustration and anger in the comments on Retraction Watch.

The case of Dr. Hans Hellinga at Duke University is an example where publication by another group was able to correct the research record, but the student subsequently was alleged to have committed misconduct by her own PI. Hellinga's group claimed that it had designed a modification of an inactive protein that converted it into a highly active enzyme. Dr. John Richard of the University of Buffalo (SUNY Buffalo) discovered an error in the observations reported by Hellinga's group in Science [357]. Richard's group attempted to reproduce the experiment but discovered that the observed activity of the enzyme was due to a contaminant. Dr. Hellinga realized his error and did the honorable thing by retracting two high profile papers. The matter would likely have ended there as an embarrassment, but with no ethical implications, if Dr. Hellinga had not alleged that his graduate student, Mary Dwyer, had committed research misconduct. That allegation led to an investigation, which exonerated the student. The article in *Nature News* on this case suggests that the graduate student attempted to communicate the experimental issues and doubts to Dr. Hellinga. The adjudication of this type of internal problem in a research group is crucial to the entire issue of self-deception since there is a distinct possibility that a subordinate learns or is instructed to implement flawed approaches from the PI. Dr. Dwyer was able to defend herself, but the numerous cases cited previously show that she is the exception. She had documented her work and had a credible case that she had informed Dr. Hellinga that more control experiments were needed. One can understand why this case became a *Nature News* story since Dr. Dwyer would need to explain to any future employer or research group why her major professor did not support her. While the resolution to these difficult ethical problems is an important current issue for the scientific community, the current research misconduct system has failed to consistently address the problem of research group relations and how these should be handled in a research misconduct proceeding.

When it was discovered that a Columbia University chemistry PhD student, Bengü Sezen, had manufactured data in several publications that were the basis for awarding tenure to her supervisor, the student was immediately expelled [358]. Since Columbia University refused to make any comment on the case, no one knows if there were internal sanctions, but from all outward appearances there were no consequences for her mentor. It is known that three graduate students who overlapped with Ms. Sezen and that these students reported difficulty in repeating her work. Their criticism was apparently not well received by the mentor since all three left the university without obtaining a degree. Reporter William Schulz of *C&E News* attempted to understand the history behind these events [359]. He questioned why Columbia ordered its staff not to speak to journalists about any aspects of the unfortunate affair. In this case, the "bad apple" defense seems particularly weak since students suffered by coming forward and attempting to show that there were significant discrepancies. One would hope that a professor and mentor would listen very carefully when provided evidence of the type that is described in Schulz's article.

6.10 The Scientific Establishment in Denial

For more than 35 years, the literature on ethics in science has been documenting a pervasive problem, which has roots in the educational system itself [8]. The problem of academic fraud appears to have been easier to recognize in certain other countries. Articles in *Science* and *Nature* have pointed to the seriousness of fraud in China between 2005 and 2010 [360, 361]. For example, the fact that 70 papers were retracted from a crystallography journal in 2009 has been reported repeatedly in *Nature* and *Science* without emphasizing that all the papers were produced by a single research group at a relatively obscure university, Jinggangshan University in Jiangxi province [98, 362, 363]. The authors, Zhong Hua and Liu Tao, were fired immediately after the falsification was discovered. The urologist Chuanguo Xiao was alleged to have falsified data, lost his academic job and then attacked the anti-fraud activist Shimin Fang with pepper spray and a hammer in 2010 [362]. These cases show that there are consequences for faculty in China when scientific misconduct is proven. However, the tone of the articles written in Western magazines would lead one to believe that there is a lax attitude towards the epidemic of falsification in China. For example, articles have cited a study that claims that one third of all Chinese scientists have admitted to ethical violations [363]. While there clearly are problems in China, are we in the West really any different? In fact, with a similar proportion of scientists in the United States have admitted to such transgressions [301, 364]. Every country that has carried out such anonymous questionnaires has found alarmingly high numbers of scientists who admit that they have some knowledge or observation of improper procedures in research [365]. Since the publication of high-profile incidents in China, the pressure on Chinese scientists is definitely greater than on Western scientists [98]. For example, the president of Zhejiang University called for a zero-tolerance policy towards research falsification and fabrication [366]. While it is easy to point the finger elsewhere, it is more difficult to confront the issue at home. There is good reason to believe that the failure to properly adjudicate research misconduct is a global phenomenon. It is indeed possible that the reason the problem appears worse in China is simply because the Chinese authorities have been more public about their actions than their Western counterparts. The secrecy and legal threats in the United States have meant that many U.S. professors who have falsified or fabricated research were able to move to other universities or research institutes.

The problem of misrepresentation of science is clearly a global phenomenon. Two German professors, who were rising stars in the field of cancer research, committed the largest research fraud in the history of the German university system starting in 1992. Friedhelm Herrmann and Marion Brach falsified dozens of publications over a period of 10 years to further their careers and obtain positions at top universities [367]. The students had suspected the fraud for some time, but were afraid to come forward because of the consequences for junior researchers. During much of the 1980s Great Britain was free of the scandals that had rocked science in the United States. If there was any complacency in Great Britain it was shattered in 1994 when the case of Malcom Pearce came to light [368]. Dr. Pearce, a gynecologist had published a paper claiming that he had successfully treated a woman with an ectopic pregnancy and that she had subsequently given birth to a healthy baby girl. That turned out to be false and also revealed other fabrications. The case of Dr. Andrew Wakefield in Great Britain highlights the financial motive for research fraud as well. In 1988, Dr. Wakefield published a study in the British Medical Journal Lancet showing a link between measles-mumps-rubella (MMR) vaccine and autism. Wakefield's study on 12 children had an unacceptably small sample size, did not have proper ethical approvals and was not reproducible [293]. Evidence suggests that Wakefield manipulated data for financial gain since he was an expert witness in lawsuits against vaccine manufacturers [369]. Sadly, all too many people believed the results of this fraudulent research. Fifteen years later there have been epidemics of measles in Wales and in pockets throughout the United States because a small, but significant, percentage of the population has not been vaccinated due to parent's fears of autism based on Wakefield's claims [370]. Meanwhile, despite losing his medical license, Dr. Wakefield is alleged to have benefited from all this by receiving a large payment in return for his involvement in a legal case that was being mounted against the vaccine manufacturer [372]. This case underscores a crucial reason for maintaining an accurate research record. The public can seize on ideas in the scientific literature and misinformation can propagate for years.

Among the most egregious frauds in the history of academic research was committed by Dr. Ragner Rylander of the University of Geneva, who had had a secret contract with the cigarette manufacturer Phillip Morris. For nearly 30 years he used bogus scientific research results to minimize the dangers of tobacco smoke [223]. In 2001 the University of Geneva, exonerated Dr. Rylander until a civil court in Geneva had accepted a lawsuit against Rylander when the full extent of the financial ties with Phillip Morris became clear [224]. This case demonstrates the unwillingness of a university to confront the ethical issues when an established and well-funded research scientist is being investigated. While the University of Geneva declared that it had learned its lesson, the real question is whether any university can conduct the necessary self-examination when the financial balance sheet is affected by the outcome of an investigation.

The claim of cold fusion made in 1989 by Dr. Stan Pons and Dr. Martin Fleischmann has not been directly called fraud, but it clearly is unsubstantiated science that arose from a self-deception [372]. Suspect claims of nuclear fusion surfaced again after 2000 when Dr. Taleyarkhan of Purdue University claimed that nuclear fusion had been observed in ultrasonically created bubbles. to gain credibility for the idea, Dr. Taleyarkhan was alleged to have misrepresented a published paper by placing former post-doc as the author (and not putting himself on the author list) so that it would serve as independent confirmation. The claims were investigated and Dr. Taleyarkhan was exonerated by a committee at Purdue University [22]. However, the Subcommittee on Oversight of the Committee of Space, Science and Technology (SST) of the U.S. Congress weighed in by asking the Office of Naval Research (ONR) to reopen the case [373]. The ONR in turn asked Purdue University to reinvestigate. Subsequently the matter was reinvestigated leading to a finding of research misconduct [374]. Dr. Taleyarkhan continues to work at Purdue University.

Two Duke University medical researchers, Dr. Anil Potti and Dr. Joseph Nevins, founded the company Oncogenomics based on their cancer research that had been published in Lancet Oncology in 2007 [375]. The company changed its name to CancerGuide Diagnostics and raised millions of dollars in capital to advance the concept of a personalized medicine based on genetic analysis. Although this general concept may have promise, Dr. Potti quickly moved forward into clinical trials on the basis of data that were later retracted by the editors of Lancet Oncology. He administered drug cocktails that were not approved for use in cancer patients. The statistical models used to justify these treatments were questioned by Mr. Perez, a medical student working in their laboratory. Mr. Perez took his concerns to the Duke administration. He was discouraged from pursuing allegations by Dr. Nevins, who was Dr. Potti's mentor and business partner. In the end Mr. Perez distanced himself from the group and repeated a year of medical school he had lost while working in Potti's laboratory [376]. Then, Drs. Baggerly and Coombes began to review the data sets and found inconsistencies [376]. Yet, Duke University conducted an internal review and concluded that the methods were in order. Subsequently, the clinical trials were halted and Potti's data was subjected to a National Cancer Institute (NCI) review. When Potti presented the data to the NCI panel no errors were found. In an interview with 60 min years later, Nevins admitted that it was a mistake to permit Potti to select the data to present to NCI. The treatments based on a faulty statistical model continued for several more years with over 100 terminally ill patients in the clinical trial. Finally, Nevins himself began to look at the data and discovered that it had been manipulated. At least nine publications have been retracted by Duke University. Nevins is still at Duke University and Potti moved to South Carolina where he continues to practice oncology.

X-ray crystallography data deposited in the Protein Data Bank provide a record that can be used to reconstruct fraudulent structures. The telltale signs of falsification can be quite subtle and it is work for a sleuth to find evidence of falsified structures. Rupp has described the discovery of a forged structure of birch pollen allergen in didactic fashion [377].Although not forgeries, Stanfield, Pozharski and Rupp have exposed structures of drugs or small molecule ligands bound to proteins with little or no electron density to support the claimed binding [172, 378]. As in other areas of science the range of ethical transgressions is evident, but in x-ray crystallography the nature of the data provides extensive evidence on the intent and methods of those who have published sloppy science.

Although the premeditated frauds are perhaps the most disturbing, self-deception can lead to fraud without intent from the outset. Serious self-deceptions can be as harmful to the scientific literature as intentional frauds. To the next scientist trying to make sense of the published result it does not matter what the motive was. There is a range of possibilities that go beyond the "bad apple" theory, which is the notion that scientific fraud results from deranged individuals or con men [94]. We must consider the strong possibility that there are causes rooted in the culture of modern science, which is strongly affected by the failure to properly teach research ethics while simultaneously tolerating the commercialization of research. While both have received recognition as problems, the combination of this culture with administrative culture of confidentiality to protect the institution is toxic. Although the thesis of this book is the university conflict of interest is a central aspect of the negative ethical climate ushered in by federal research misconduct regulations, scientists must come to terms with the need to educate themselves on this issue.

After listening to the denials in a Congressional Hearing, Al Gore, who headed the 1981 Congressional Committee hearing on Ethics, said "I cannot avoid the conclusion that one reason for the persistence of this type of problem is the reluctance of people high in the science field to take these matters very seriously" [88]. Scientists were forced to accept the action of Congress to mandate a research misconduct procedure. In the late 1990s a policy was formulated in the U.S. to investigate and punish falsification, fabrication, and plagiarism, which were collectively called "research misconduct". Unfortunately, the existence of research misconduct system has provided another excuse for the leaders of the scientific community to deny that a problem exists under the assumption that it is being adequately managed by universities. It is odd that American scientists who live in a country whose government is founded on the principle of checks and balances have a difficult time realizing the need for the same approach towards ethics within the scientific community. Instead, many scientists repeat the idea that science is self-correcting because papers will eventually be published in the scientific literature that will correct the record.

Chapter 7
University Administration of Scientific Ethics

> *"Be not ashamed of mistakes and thus make them crimes."*
>
> *Confucius*

The role of the university administration is crucial to the adjudication of research misconduct because of the trust placed in universities by federal regulations. Nonetheless, those regulations provide no compelling reason for a university to take on the challenge of going through the evidence, consulting experts, and determining the details of how a scientific project may have gone awry. The regulation gives the university the unfunded mandate to investigate and judge its own faculty, but also gives the university the right to declare that there is no need for an investigation. In the first decade after the creation of the ORI there were numerous lawsuits by individuals and the U.S. Government against universities under the False Claims Act, even when universities self-reported the misconduct and took action [379]. Under the False Claims Act a whistleblower can receive a significant percentage of fraudulently spent grant money and universities must pay the whistleblower as well as repay the federal agency that granted the funds.

Although the liberal interpretation of the False Claims Act lasted only a few years, the precedent established during that period gave universities reason to be reticent to report allegations unless absolutely necessary. Universities felt the enormous burden of the cost of the investigation compounded with the loss of revenue. Universities realized that if a RIO determines that an allegation is frivolous, the allegation can be dismissed without any record (outside of a file in the RIO's office). Alternatively, at the inquiry stage, an inquiry committee of three faculty members chosen by the RIO may excuse the matter. A report must be filed with the funding agency, but as long as it contains the necessary language regarding the convening of a committee, the report could state that the evidence is insufficient to warrant any further investigation and no questions will be asked in most cases.

At the conclusion of an investigation, federal regulations under the FOIA require the OIGs or ORI to release the report and other information about the case in redacted form, i.e. with black markings over names and other identifiers. In

S. Franzen, *University Responsibility for the Adjudication of Research Misconduct*, https://doi.org/10.1007/978-3-030-68063-3_7

practice, the release of specific documents becomes a legal question, but there are government lawyers in a FOIA office who will redact documents so that they accord with the requirements of the Privacy Act. The spirit of sunshine laws in each state or the FOIA for the federal agencies is that the institutional decision-making process and course of action is in the public domain, but the individual actors should remain anonymous. If one wishes to obtain any information pertaining to case, the greatest difficulty is in asking for the documents in the correct way. After persistent questioning federal lawyers and legal aides in the various FOIA offices will provide assistance in many cases by giving clues to the uninitiated permitting one to properly request a document. Although they are bound by many rules, including the confidentiality regulations, those employees see their mandate as upholding freedom of information. Unfortunately, universities often do not release documents even when the federal guidelines recommend it. Therefore, there is a paucity of information available for the study of the problems that confront the scientific community in the adjudication of research misconduct. Thus, when a case is made available it can serve to understand the types of problems that are confronted on a regular basis in hundreds of universities throughout the world. This chapter takes the reader behind the scenes to examine facts that were not evident in the research articles discussed in Chap. 1. The descriptions in this chapter based on the facts of the hexagon case provide a rare look behind the scenes at how investigations are conducted by universities and the role played by the federal agencies in the hexagon case. This information is public but it would be nearly impossible for an outsider to obtain the information. This chapter presents essential evidence in support of the hypothesis that university conflict of interest prevents proper adjudication of research misconduct allegations.

7.1 The Discovery of Discrepancies Between Laboratory Results and Published Statements

The first evidence of falsification of scientific data in the study of palladium hexagons made by RNA was presented at University 1 in December 2005 when a graduate student in our collaborative team first showed conclusive data that the hexagons were not composed of palladium metal. From that point to September 2006, a graduate student and post-doc led me to the conclusion that the data had been misrepresented in the publications. In the beginning, I did not suspect anything was wrong, because I accepted the respondent's statement that we needed to study freshly made samples to be sure that any conclusion reached was representative of the samples presented in the publications. Over the course of next few months, the research done by a graduate student and a post-doc in our collaborative group demonstrated inconsistencies and worse, misrepresentations in the published accounts of the palladium particles. Preparation of new samples did not resolve the issues. By February 2006 I had become concerned that something was seriously wrong. By this time, my laboratory had been part of the W.M. Keck Center for RNA-Mediated Materials Synthesis for over 1 year. Over the course of the ensuing months, a graduate student and post-doc systematically went through possible explanations. I had seen apparently inexplicable discrepancies before in my career, and I knew that it sometimes took months to figure out the explanation for experiments that gave unexpected results. Moreover, in my experience there had always been a good explanation for apparent discrepancies once we understood the problem more deeply. Unfortunately, the hexagon case was different from anything I had experienced in 20 years of working in a laboratory. The bad news continued to arrive, and it was not just the composition of the hexagons that was incorrect. During 2006 the graduate student and post-doc produced six internal research reports documenting various experiments that were inconsistent with the claims of the publications. A professor of Materials Science and Engineering who also mentored the graduate student became important in the adjudication of the hexagon case. He not only supported the graduate student's interpretation, but he had done independent work on samples prepared by two other graduate students working with the respondents. As experts in the methods of electron microscopy the professor showed that the composition and structure of particles were completely misrepresented in two separate projects being pursued by the two respondents. Of course, we provided the reports to the respondents.

It is awkward to be showing colleagues that you are proving that their work is wrong, but also misrepresented in the articles. The series of proofs culminated in an experiment under a microscope. The graduate student made large hexagons and then showed that these large hexagons could be dissolved in pure THF (an organic solvent that is miscible with water). This observation would be analogous to a gold ring dissolving in rubbing alcohol: it simply does not happen. The fact that we could see the hexagons dissolve in a few seconds under the microscope was quite convincing as a visual demonstration that the hexagons were not metallic. These simple experiments invalidated the entire premise of a new technology, in which RNA was a mediator that created new inorganic materials. As these discoveries were made, our collaborators avoided discussions or meetings we held to discuss and analyze the findings. Finally, they threatened us in various ways. The graduate student was intimidated by one of the respondents, called incompetent, and ordered to cease work on the project. The post-doc was excluded from respondent's laboratories where she had worked for years. When I told the two professors that I thought we should publish a Technical Comment in *Science* together, they threatened me by saying that I had violated intellectual property law and would be sued by a company that they claimed was somehow involved in the work. This was a surprise to me. I had no relation with that company and there were no documents in our grant proposal that gave that company any rights to the samples we had studied. Without my knowledge, a process of retaliation had already begun behind the scenes. I was aware of the verbal threats, but it never occurred to me that these were followed up with allegations in writing against me that would then be taken seriously by the university. I assumed that I would be informed of any investigation if an allegation against me were made. All my assumptions about a university administration would handle such a matter turned out to be incorrect.

By November 2006 my research group had conclusive evidence that the result published in *Science* was based on false statements and that there were numerous attempts to cover up the facts. In a meeting with my Department Chair and Research Dean in October 2016, one of the respondents repeated the claim that all the work we had done on the project was owned by a company in State 2. The respondents threatened a lawsuit for intellectual property violations if we were to submit a Technical Comment to *Science*. While this did not seem like a credible legal threat to me, the University 1 administration prevented us from submitting any publication and forced me to work with the University 1 RIO to obtain permission to submit any publications. I had accepted a considerable amount of funding and was denied any opportunity to write a correction by my own university, I felt that ethically I had no choice but to tell the RIO what I knew. To avoid humiliating my former colleagues, I asked the RIO if we could not find an impartial arbiter or academic ombudsman who would help to find a resolution. There was no formal process for mediation in such cases at University 1. A senior faculty member in our department tried to mediate, but that quickly broke down.

When I met with the RIO to discuss the steps we needed to take to establish that the samples were common property and that we had every right to publish our findings, the RIO informed me that there were sufficient grounds for a research misconduct investigation. No matter what I called the discrepancies that I described, the RIO explained that federal regulations required him to call them allegations of research misconduct. I was required to provide a written description of the scientific misrepresentations to the RIO in order to convince General Counsel of University 1 that we should have the right to publish our findings, I wrote a detailed account of what we had observed and how it contradicted the published account in *Science*. I used the word "misrepresentation", but not "falsification". In accepting what I had written, the RIO explained that he considered my description to be an allegation. The RIO insisted that by regulation he was *required* to initiate a research misconduct inquiry, unless the "allegations" were deemed frivolous or made in bad faith. The legal threat made by the respondents had forced the issue since it blocked the avenue of correcting the research record by publication for many months, and their legal threat virtually guaranteed that there would be an inquiry. Indeed, the RIO did give me the right to withdraw my "allegations", but then it was not clear whether we would have the right to publish findings to correct the record. By this point, I had to concede that misrepresentations and actions to cover up the truth were evidence of falsification. At this time, I did not yet understand that the definition of research misconduct in federal regulation 45 CFR § 689.1 required proof of "recklessness" or "intent". There was a large chasm between what I thought the allegations implied and the actual process that followed. I deeply regretted that things had reached the point where I filed allegations, but I knew that what had I had observed from the respondents' laboratory did not agree with what they had published. My only scientific reservation was that I had not yet seen the laboratory notebooks. I assumed I would see them as part of the process, but that turned out to incorrect.

7.2 The Research Misconduct Investigation into the Hexagon Case

The research misconduct investigation of the hexagon case has been described in newspaper articles [30, 32–35, 38], *Retraction Watch* [14, 380, 381], scientific magazines, such as *Nature News* [41], *C&E News* [36] and *Science Insider* [37], in the final NSF report [42] and in court documents filed in Superior Court of Wake County, North Carolina (16-CVS-12363) [382]. Since many documents of the hexagon case are in the public domain, this case is one example where we can study the effect of university conflict of interest on a research misconduct investigation. Even so, the university has refused to release many public records pertinent to the case, particularly the ream of legal threats that they received from the respondent's lawyers. The respondents sent allegations of my wrongdoing to many offices in the administration, to editors and to federal agencies. They had a lawyer for procedural issues in the university and the investigative process and a second lawyer who made threats concerning perceived intellectual property matters. The actions taken by University 1 to investigate me and the threats made by those lawyers were kept secret from me. The only way to obtain the documents sent by those lawyers and the allegations made against me would be to pursue a public records case to conclusion, and most likely through an appeals process that would take many years. Those documents might explain why it took 9 years for two universities and the NSF OIG and two other federal agencies to adjudicate a scientifically trivial case of falsification. The two respondents conducted the initial research at University 1 and then moved to University 2 to continue the research with significant additional financial support from that university. Once University 2 had paid for their move as full professors that investment combined with previous investments in related technologies by one of the professors became a reason to protect the two professors. Unfortunately, that defense involved legal attacks on the credibility and integrity of the whistleblower. Discrediting the whistleblower is a common strategy, and anyone who reports falsification to a university RIO or a federal agency should be prepared for a counterattack. Indeed, I was expecting some reprisal, but the extent and duration of attack on my credibility and integrity was far beyond anything I could have imagined. University 1 did not defend me in any way, but rather joined in the attack by withholding information from me about a federal fraud investigation, suggesting to federal investigators that I was guilty of fraud, investigating me in secret, making a judgment about my integrity and scientific ability and finally by initiating a federal inquiry into my right to use data obtained in the collaboration.

The three authors of the 2004 *Science* publication consisted of one graduate student and the two professors. The University 1 RIO and NSF OIG both designated the graduate student as a respondent, because she was the first author of the *Science* publication. I insisted from the beginning that the graduate student was misled and was a witness rather than a respondent. If mentors continually tell a graduate student that he/she should interpret the findings a certain way, the student is not in a position to contradict that instruction. Furthermore, if mentors take control of the process of writing an article because of their greater experience, the student may have little say in the writing of the article. I believe that the entire matter could have been rapidly resolved if the graduate student had been treated as a witness, rather than as a respondent. In 2008, the graduate student freely admitted to collaborator, Dr. De Yoreo, that she had not used the solvent mixture reported in *Scanning* and, instead, had used 50% THF/50% water. This simple point, which was denied by the former University 1 professors to the end, would have been resolved by her testimony as a witness, rather than a respondent. Instead, in her interview as a respondent, she was evasive in her interview with the University 1 committee saying as little about the THF percentage as she could under interrogation. Since she used 50% THF/50% water in the experiments and the two professors wrote that she used 5% THF/95% water in the article that they called their "correction", she would have been conflicted about how to answer questions about what solvent conditions were actually used. Telling the truth that she knew would have undermined their defense. They were asking her to testify against herself when she was not the one who committed the offense. On the other hand, she had not written the publications and she would have needed to stand up to two professors to correct what they wrote in any discussions prior to publication. Under the rules implemented by the university administration, any admission on her part that the published values were falsified would have been evidence that pointed towards her guilt as well. How could she explain to the committee that she had no control over the way her mentors described her work in the *Science* paper? Despite this fact, the University 1 inquiry report shows that she readily admitted that the solvent conditions consisted of "much higher than 10%" THF when asked a direct question by the committee, although the committee never pursued this point. This is another indication that she was disposed to be truthful. The committee failed to pursue the matter because the committee did not see the laboratory notebook where the actual values of 50% THF/50% water were given for every experiment except the very first one, which was a failed experiment.

University 1 held the original patent application (PCT/US2004/029480) [31] and had supported the initial phase of the research but stood to gain nothing financially once the respondents moved to University 2. The RNA mediation technology claimed in the *Science* paper was intended as the basis for a research center and for a company at University 2. Yet, University 1 clearly had jurisdiction over the first three articles [29, 44, 45] because the research had been conducted at University 1. University 1 was left "holding the bag" while the respondents left with their government grants, equipment, and any future commercialization opportunities. Rather than recognize the extent to which they had been used, administrators from University 1 at the highest level sought to protect the respondents, perhaps out a fear of loss of prestige or fear of an intellectual property lawsuit, which the respondents repeatedly threatened. University 2 attempted to wrest control of the entire matter from University 1 because University 2 had made a significant financial investment in the technology by hiring the two professors and providing them with "start-up packages" totalling more than $1 million to transfer their research. Most of the original grant money had either been spent or moved to University 2 before the research misconduct investigation began. It is quite clear from subsequent events that University 2 intended to end the investigation if they possibly could. University 2 had a vested interest in protecting its investment by stopping the inquiry into falsification. Only one administrator at University 1 acted on principle. The RIO of University 1 did not permit University 2 to wrest control of the investigation from him.

Appendix A provides a timeline to show how the debate in scientific journals proceeded in parallel with the research misconduct investigation. University 1 initiated a research misconduct inquiry against the two professors under the jurisdiction of the NSF since one of the grants that funded the research had originated in that funding agency. By summer 2007, the University 1 investigation committee (IC) found sufficient evidence of falsification for a full investigation. The University 1 IC of five professors continued its work for another year and submitted a report to NSF in June 2008 that concluded that there had been falsification of data and that the respondents had acted outside the norms of acceptable conduct established by the scientific community [32]. However, the University 1 report stopped short of a finding of research misconduct. The five professors were divided on the question of whether there had been intent or recklessness. Without a finding of either intent or recklessness the conclusion is not called research misconduct, even when there is agreement that data were falsified. As an example of their deliberations concerning falsification, which spanned nearly ten pages in the report, they wrote:

> "The Committee determined that the Respondents' use of the statement "A combination of scanning electron microscopy and electron diffraction showed that the hexagonal particles were crystalline Pd" (pg 851 of the Science paper) is a falsification, because it misrepresents the data on which that conclusion is based, and therefore prevents the knowledgeable reader from reaching his or her own conclusions or reinterpreting the claims of the Science paper in light of additional evidence. By definitively claiming metal-metal bonds and crystalline palladium without disclosing primary data, the knowledgeable reader would assume that these claims were beyond scientific dispute at the time the Science paper was submitted. They very much were not."
>
> – University 1 Investigation Committee Report [33]

Despite the finding that data were falsified outside the norms acceptable to the scientific community, the conclusion of the University 1 report was that there was insufficient evidence of intent or recklessness to warrant a finding of research misconduct. The matter did not rest there. By regulation, the federal agency responsible for the funding must accept the conclusions of the report. In this case, the NSF OIG did not accept the findings by the University 1 committee. Because of the shortcomings in procedure, such as the failure to consult the primary evidence in the laboratory notebooks, the NSF OIG conducted their own investigation from 2008–2013. During this entire period confidentiality regulations required that no public statement be made, not even a notification that an investigation was underway. In September 2013, nearly 7 years after the allegations were filed, the OIG sent a recommendation of research misconduct to the Deputy Director of the NSF, to the U.S. Congress and ultimately in May 2014 to University 1. In the Semi-Annual Report to Congress in the autumn of 2013 the NSF OIG described their conclusion that the authors of the *Science* paper had recklessly falsified data and recommended a finding of research misconduct, which would carry a punishment of debarment from NSF funding for 3 years. The NSF OIG also recommended retraction of the *Science* paper [33, 38].

According to the regulation (45 CFR 689.9) the Deputy Director must either approve or reject the finding of research misconduct within 120 days. It took much longer. Two years later, in September 2015, the NSF Deputy Director concluded in agreement with the University 1 report, that the respondents had committed "falsification" outside the norms of the academic community, but not research misconduct [36]. At first reading the lack of a finding of research misconduct appears to be a rejection of the NSF OIG finding, On closer inspection, the letter of reprimand the NSF referred to the NSF Charter of 1950 and debarred the respondents from NSF funding for life for a "failure to report" their results accurately to the NSF Director. In fact, the punishment for this sanction was stronger than that for research misconduct, which is a debarment for a period of 3 years. The letter of reprimand stated that the conviction would be removed if the respondents would "correct" their original misleading statements in the *Science* paper. This is also what University 1 had requested from the respondents 7 years prior. According to the final NSF report, up to that point the "corrections" had been nothing more than restatements of the previously published misrepresentations [42]. At the conclusion of the NSF process the respondents sent yet another "correction" to the Editor-in-Chief of *Science*. Based on the account in *ScienceInsider* [37], Editor-in-Chief, Marcia McNutt recognized that the respondents had not submitted an actual correction. By this time many journals had accepted "corrections" by the respondents, but none of these had accurately described the "errors" made in reporting the conditions and structural data. Based on consultation with the NSF Director, the Editor-in-Chief took unilateral action to retract the *Science* paper. On the other hand, the remaining falsified data papers in *JACS, Langmuir, Scanning* and the *Journal of Materials Chemistry* have neither been corrected nor retracted [44, 45, 65, 383].

The 41-page NSF Final Report provides details of each publication and witness testimony was described [42]. The initial section on the University 1 investigation discussed the conclusion by the University 1 IC. In Sect. 7.2 the OIG gave its reasons for opening its own investigation. The OIG found that the University 1 committee had not appropriately consulted the primary evidence in the laboratory notebooks. The role played by the laboratory notebooks is described in an excellent newspaper article entitled "In a notebook at University 1. A smoking gun" [30]. The OIG investigation included interviews of witnesses and comparisons with the primary data in the laboratory notebook. In the section on their investigation they revisited the issue of the composition of the hexagons, the reported solvent conditions and the lack of control experiments. In each case they found a lack of agreement between the primary record and the publications.

At the end of Sect. 7.2 the OIG examined two inquiries conducted by University 2. The first inquiry by University 2 was opened during the inquiry phase at University 1. Although University 1 had jurisdiction, University 2 proceeded with their own inquiry. The first University 2 inquiry quickly concluded with an exoneration of the respondents, or at least of one them. No evidence was sequestered, and no witnesses were interviewed. The assurances by the respondents that the entire matter was a dispute with a dissatisfied former collaborator were sufficient to close the case. After the University 1 investigation closed in June 2008, a second University 2 inquiry was held during the fall of 2008 using only an unauthorized copy of the University 1 investigation report as "evidence". At the conclusion of their brief deliberations the University 2 Standing Committee wrote:

> "It was unanimously agreed that the first two allegations were not significantly different from the allegations already investigated by University 1. Given the thorough evaluation completed by the University 1 IC [Investigation Committee] and its exoneration of the two respondents of any and all charges the Inquiry Subcommittee unanimously concluded that these two allegations were unfounded."

In reexamining the statements made by the University 2 Investigation Subcommittee the NSF OIG wrote:

> "This statement is an inaccurate of summary of the findings of the University 1 IC. The University 1 IC concluded that the *Science* paper contained a falsified conclusion, and that the subject's actions in its publication were a serious deviation from accepted practices. The university adjudicator concluded that the distortion of the scientific record was sufficiently great that correction of the record was warranted despite the absence of a finding of research misconduct. Thus, we [the OIG] do not interpret University 1's finding as an "exoneration" of the subjects."

These statements were kept confidential until 2015 when the report was released. No one in the scientific community knew that the University 2 inquiry was a cover-up. The respondents used it widely to convince journal editors and other scientists that the whistleblower had been wrong and that the entire investigation had been inappropriate. University 1 was made aware of this, but took no action to defend the whistleblower.

In Sect. 7.3, entitled OIG Assessment, the OIG presented its assessment that the falsification rose to the level of research misconduct based on the repeated reckless disregard for the scientific facts of the case by the respondents. The intent and recklessness were each considered in making the determination. The report noted that the graduate student's notebooks directly contradicted the statements made in the 2004 *Science* paper. Section 7.4 (The Subject's Responses to the OIG Draft Report) gives an account of how an OIG decides on intent or recklessness. In the hexagon case, the NSF OIG found recklessness, but not purposeful intent to falsify. One of the respondents objected that recklessness is a "vague and undefined label". The OIG staff wrote that, on the contrary, recklessness is a "well-established legal standard akin to gross negligence." The objections to the NSF OIG statements by the Subjects include claims that there are "contradictions" or "errors" in the report. However, the respondents provide no evidence to support these claims. Perhaps the most telling statement is that by the younger of the two respondents who wrote "It thus appears that our conclusion is not a significant departure from common practice. Two university committees have agreed, having exonerated us of misconduct." The NSF officials wrote that "this is a mischaracterization of the conclusions of both universities."

The NSF OIG report establishes the conflict of interest of University 2 in a way that is rarely observed so clearly in a research misconduct investigation. In fact, the University 1 report conclusion was unanimous that the "falsification" was "outside the norms" established by the scientific community. The University 2 inquiry report did not accurately assess the conclusion and in stating that the allegations were unfounded, put the onus for the waste of time and resources on the whistleblower. The respondents used this report to convince editors and other scientists that they had been exonerated and that the papers we had published to correct the record were inappropriate. Neither the University 2 RIO nor the Head of the Standing Committee on Ethics wished to comment on the full OIG report when asked by the reporter in 2016 [35], although they had been willing to speak on the record 2 years prior to blame the whistleblower for bringing forward a case that they thought should have been sorted out in the journals [38]. It was an extreme irony to be attacked behind the scenes for publishing and castigated in public for failing to use the appropriate channel of publication to settle the matter.

7.3 The Deficiencies of Federal Research Misconduct Regulations

Research misconduct regulations were developed to provide a fair and expedited process for adjudicating allegations of plagiarism, falsification, or fabrication. The regulation failed in both aspects in the hexagon case. First, the time required for adjudication was unacceptably long by any standard. It took nearly 9 years from the filing of the allegations in November 2006 to the final report release on the NSF website in September 2015. By regulation, the entire matter should have been adjudicated in no more than 19 months. The NSF OIG regulations in CFR 689.4 specify that the inquiry and investigation should take 90 and 180 days, respectively. If the NSF OIG finds research misconduct as happened in this case, then according to CFR 689.9, there is a 30-day rebuttal period and 45-day processing period by NSF OIG to answer the rebuttal. Then, the process takes another 120 days for the final decision of the NSF Deputy Director. In practice, the NSF OIG will permit universities to prolong these deadlines indefinitely. Indeed, if the NSF OIG itself takes on the investigation after the university investigation has concluded then by regulation CFR 689.5 it should take no more than 90 additional days. Second, there was no consideration given to the confidentiality of the whistleblower when University 2 used the "confidential" University 1 investigation committee report without authorization to state that the allegations were "without foundation". Neither the NSF OIG nor University 1 enforced confidentiality when I pointed out to each of them that my right to privacy had been violated. Confidentiality protected the respondents since the scientific community did not know that there was an active investigation for the entire 9 years and the negative fallout for the whistleblower from counter-allegations and public statements by the respondents and university administrators will last the duration of the case. For a good faith whistleblower, even one who is a full professor, a nine-year timeline with a significant breach of confidentiality will have a significant negative impact on one's career.

As the whistleblower in the case, I came to understand the power of counter-allegations and retaliation. Counter-allegations can muddy the waters sufficiently that the administrators who are conducting the investigation may feel compelled to investigate the whistleblower with as much zeal as they investigate the respondents. Clever use of allegations such as violation of intellectual property and theft of samples can cast the original allegations in doubt. Counter-allegations of misappropriation of samples should be easy to prove or disprove. Yet the effect of counter-allegations can be to turn both parties into suspects. For the respondents, this has the desired effect of a smokescreen if the university administration fears being on the wrong side of the issue that may lead to a lawsuit. The point of having a research misconduct regulation is to mitigate these problems and hopefully to provide an efficient and speedy resolution. Instead, the current system permits universities to take the easiest path, which is to attempt to bury the matter as quickly as possible. In the hexagon case, I became a suspect because of the allegation that I had used samples and data without authorization. The allegation was absurd. I had helped to raise a significant portion of the money and I signed my name to two grants that made me legally responsible (one as principal investigator and the other as a co-principal investigator). The university knew this, of course, since they managed the grants. Yet University 1 pretended that they could not be sure whether I had stolen samples and data from colleagues as part of a vendetta against them. These decisions and their consequences were known to no one but me and I was advised that I would be punished if I let anyone know how the university was managing the investigation. I advocate for a process of open discussion and scientific inquiry as essential aspects of any ethics investigation.

The current system for adjudication of research misconduct is not based on science and there is no pretense that it should be. There is no requirement for those making the decisions in the university, including the investigation committee itself, to have any understanding of the interpretation of scientific data or how someone might misrepresent those data. Since the committee deliberations are strictly confidential, there is no outside consultation with experts unless the committee decides it is necessary. There usually is no budget for such a consultation so the committee is less likely to be able to request expert opinions. The opinions usually come from professors at the university who do not know the facts of the case (since these are confidential). Since all proceedings are confidential, the only way for the whistleblower to know whether the committee used any scientific criteria in their decision is from reading the reports at the conclusion of the investigation after the fact. The regulation does not say definitively that the whistleblower even has a right to see those reports. Thus, the right to inspect the deliberations may become an administrative or even a legal battle under current regulation. Initially, University 1 rules were interpreted to give me the right to the inquiry and investigation reports. Then later, when I criticized the University 1 administration for its failure to notify journal editors (confidentially) of the outcome of the University 1 investigation,

University 1 General Counsel's office alleged that I had obtained the investigation committee report by a "unauthorized channel". Fortunately, I saved the emails showing that I had been given the report by the RIO in accord with University 1 Rules and Regulations. Even though receipt of the interim and final report by the whistleblower was in accord with university policy, every time I attempted to discuss any aspect of the process with University 1 General Counsel, I was told that the report was confidential and that I had no right to discuss it, even with General Counsel's office.

7.4 Legal Threats and Administrative Actions to Prevent Correction of Journal Articles

Pressures on University 1 by lawyers working for the company associated with the respondents and from university administrators at University 2 influenced the institutional decision by University 1 not to notify the journal editors of the finding of falsification. The University 2 administration did not keep confidential the fact that they thought that the allegations were baseless. Reporter Eugenie Reich wrote an article for Nature News entitled "Acrimony over Nanoconstruction" in which she wrote,

> "After receiving a complaint from Franzen, [University 2] held its own inquiry into the matter. Although it relied on the outcome of the University 1 investigation, the inquiry concluded that the allegations about the *Science* paper were "wholly lacking in substance". University 2's research-integrity official, [...] says that the university has run two further inquiries into allegations from Franzen. Neither was upheld, and [that official] thinks the issue is a scientific disagreement being wrongly pursued through the research misconduct system." [41]

In May 2009 I met with the RIO and with the Vice Chancellor for Research of University 1 to discuss a university policy stating that journal editors should be notified in cases where a university investigation finds falsification (University 1 Regulation 10.00.02 on Research Misconduct paragraph 6.6.2). This was our only meeting. Since the University 1 investigation concluded that there was falsification of data in the hexagon case in 2008 the three of us agreed that it would be appropriate to notify the journal editors of the finding. When the respondents were provided with a courtesy copy of the proposed letter to the editors, their response was a six-page letter from an attorney representing the company in State 2 that the respondents had repeatedly claimed owned the data as IP. This claim was false since by 2009 there the IP related to the hexagon "technology" had been permitted to lapse due to non-payment of patent fees by University 1. Nonetheless, a threatening letter from an out-of-state law firm ended the idea of contacting the journal editors. I do not know any of the details since that letter has been kept confidential up to the present. Many letters had been received from that attorney over the years and they had a negative effect on the attitude of the university towards me and towards any resolution that involved correcting the scientific record. I was never permitted to see any of those letters. A public records lawsuit ended in mediation, but without release of any documents and without a plan to correct the research record.

The pressure on the whistleblower did not end with the legal threats and public statements calling my integrity and competence in question. I had been investigated in secret by University 1 and by the fraud division of the NSF OIG, both because of baseless allegations made by the respondents, but the most significant attack came after the NSF OIG had made a finding of research misconduct against the respondents. Instead of recognition that I had acted as a good faith whistleblower, in February 2015 I was given a dossier by the University 1 RIO that informed me that I was the subject of a research misconduct inquiry concerning a paper I had written to correct the research record [25]. In August 2016 after an 18-month inquiry, in which the committee gathered no evidence and interviewed one witness, University 1 General Counsel closed the case and admitted that the allegations had "retaliatory animus".

Under the current system, there is only one oversight body, the House Subcommittee on Investigations and Oversight. While that Subcommittee did write a strongly worded letter to University 1 on July 15, 2014, University 1 has never followed its rule (10.00.02 Rules and Regulations 6.6.2), which requires notification of affected parties in cases where a university committee finds that data have been falsified. The response by the University 1 Chancellor made several excuses for failing to notify the editors. It is true that as of 2014 the Deputy Director of NSF had not made a final ruling about research misconduct. The letter of reprimand from NSF stated:

> "With respect to the accepted research practices, what the investigative authorities found as a matter of fact was an avoidance of protocols, a failure to meet expected scientific standards, a lack of expertise or training in the field of inquiry, poor oversight of less experienced scientific team members, and the misrepresentation of data on which a conclusion was based. In short, they uncovered what most in the scientific populace would deem an absence of care, if not sloppiness, and most certainly a departure from accepted practices of the relevant research community."
>
> – Case Closeout A06110054 Letter of Reprimand, May 14, 2015

The NSF directorate did not find recklessness based on a preponderance of the evidence. They wrote:

> "In this case, NSF has identified significant findings that you and your co-authors failed to disclose including: 1) a clear and consistent description of the composition of the [hexagons] used in Publication #1; and 2) the fact that data was not properly [indexed] prior to Publication # 1. These omissions were as significant to the published research record in question as the significant findings that you did prepare and submit for publication."
>
> "Accordingly, consistent with§ 18620-3 of the NSF Act, I have determined that from the date that this action becomes final you shall be ineligible for a future award under any NSF supporting program or activity."
>
> – Case Closeout A06110054 Letter of Reprimand, May 14, 2015

The fact that this is not a finding of research misconduct is a statement regarding the weakness of current regulations. When respondents have sufficient legal resources, 45 CFR § 689.10 permits an appeal that has a high likelihood of overturning any agency decision. The NSF knows this. The NSF was forced to find another avenue for sanction outside the current research misconduct regulations if it wanted to ensure that the action taken by the NSF Deputy Director was final and could not be appealed.

7.5 Public Records According to the Freedom of Information Act (FOIA)

While this section is highly specific concerning details of state and federal sunshine laws, it is crucial to understand both the barriers preventing the release of documents and methods that can be used to obtain those documents. Starting in August 2015 while I was subject to a research misconduct inquiry based on an allegation submitted by the respondents, I requested all the documents related to the hexagon case from the beginning in 2006. These requests were formalized as public records requests starting in 2009. The Public Records Act of the State of North Carolina, which is General Statute ("G. S.") § 132-6, played a crucial role in the adjudication of the hexagon case since it was used to obtain the laboratory notebooks of the graduate student in 2010 [30]. However, University 1 General Counsel refused to release most of the documents in the case using the justification that they were "confidential". Confidentiality consistently was used to prevent me from learning what legal actions had been taken to impede the investigation into the hexagon case or to prevent notification of journal editors of University 1's conclusion in 2008 that data were falsified. I tried to establish a dialog with General Counsel and the university administration by means of the faculty ombudsman (instituted in 2015), my department head, and many direct requests, every one of which was rebuffed for 7 years until July 2016, when I finally was able to meet with Deputy General Counsel of University 1. That was the only opportunity I had for resolution by direct negotiation. I asked for the public records that I had requested throughout inquiry against me and for reimbursement for legal expenses for my public records requests during the inquiry. These requests were refused by University 1.

After I had exhausted every other avenue to address the grievance caused by the research misconduct inquiry from 2015 to 2016, I filed a public records lawsuit 16-CVS-12363 in Wake County Superior Court in October 2016 [382]. Despite the denials by University 1, the investigation reports in the hexagon case are public records, as ultimately confirmed by the NSF FOIA office. The crucial point is that essentially all documents associated with a completed research misconduct investigation are public records, with the caveat that they are subject to redaction in accord with the Privacy Act. Names and personal information are blocked out. In the hexagon case, the University 1 inquiry report in 2007, the University 1 investigation report in 2008, the NSF OIG report and notification sent to Congress in 2013 and the final NSF letter of reprimand from 2015 in the hexagon case are all available because they were provided to me either directly or via a server at the NSF. However, even though the NSF made these documents available, the university maintained that these documents were "confidential".

In contradiction with the federal interpretation of confidentiality in terms of the FOIA, University 1 maintained that these and all other documents that had been produced during a research misconduct investigation were not public records. This is a contradiction since the NSF FOIA guidelines state that all documents are public records. University 1 denied this plain fact written in the regulation and claimed that it followed federal guidelines required complete confidentiality, meaning that no one could see the evidence in the case and specifically the respondent was not permitted to see the evidence against him. In this case I was the respondent.

A test case of the entire matter of what constitutes a public record arose early in the public records lawsuit. On the day that I was told of the allegations against me, I had been provided with a binder that contained the correspondence leading up to the university's decision to request a *referral letter* from the NSF. This means that University 1 requested that the NSF OIG forward formal allegations to be pursued in an inquiry. Although I had the full text of letters that proved the university had initiated the inquiry, and not the NSF, the university General Counsel and all other administrators insisted that the university was merely following the directives of the NSF in opening a research misconduct inquiry. Clearly, I planned to make reference to those letters in court. These letters had been provided to me as required by regulation, but even though that I had copies that belonged to me, the lawyers for the State Attorney General representing University 1 had moved to seal the letters, which would have prevented them from being used as evidence in open court. I had already submitted the letters to the court as evidence items by this time, but as I was about to bring that evidence to the attention of the judge I was ordered to sit down while the attorneys for the State Attorney General's office argued that the letters proving what happened should be placed under seal. That meant that I could not even mention them to the judge. The judge could read them later and draw his own conclusion, but how can anyone argue a case in court if one is forbidden from referring to the evidence. This indeed was primary evidence of retaliation. To obtain permission to use the letters publicly, I requested copies of these documents from the NSF FOIA office. Within 3 weeks, the NSF confirmed that these letters are public records and provided me with redacted copies. The redactions did not include the portions of the letters proving that University 1 had initiated the request for an inquiry. The response from the NSF OIG FOIA office on March 2, 2017 reads,

> "This letter responds to your Freedom of Information Act (FOIA, 5 U.S.C. § 552) and Privacy Act (5 U.S.C. § 552a) request, which we received on March 1, 2017, for certain documents about you in an investigative file you identified via indirect means.
>
> "Regarding your Privacy Act request, the Privacy Act gives individuals a general right of access to government agency records about themselves—but it also provides authority for agencies to exempt their law-enforcement records from that right of access. NSF has exercised this authority and exempted its OIG law enforcement records from the right of access by individuals, pursuant to 45 C.F.R. §§ 613.5(b) & -(c). Accordingly, your request for records under the Privacy Act is denied."

"If no FOIA exemptions apply to a record, however, the FOIA may require production of the record, even if it otherwise falls under the Privacy Act exemptions noted above. That circumstance exists here. Accordingly, responsive records are enclosed. Under FOIA exemptions (b)(6) and (b)(7)(C), information that constitutes an unwarranted invasion of personal privacy is exempt from disclosure. To protect the privacy interests of individuals who were not, with respect to the matter or matters described in the requested records, subject to successful criminal prosecution or recent civil legal action, and other individuals who were evidentiary witnesses in the matter, we have redacted the names, titles, and other identifying information for all such individuals other than you, pursuant to FOIA exemptions (b)(6) and (b)(7)(C)."

This letter is important because it states NSF policy regarding the release of documents pertinent to a research misconduct investigation. The significance of this letter is that *all* correspondence, reports, and evidence in a research misconduct investigation are public records, provided they are redacted consistent with the Privacy Act. Many universities, including University 1, have invoked confidentiality clause in the NSF OIG regulations and the Privacy Act to deny access to documents. For example, as their justification for sealing the correspondence, the lawyers for University 1 cited 45 CFR 689.2(h), which reads

"**(h)** Investigative and adjudicative research misconduct records maintained by the agency are exempt from public disclosure under the Freedom of Information Act (5 U.S.C. 552) and the Privacy Act (5 U.S.C. 552a) to the extent permitted by law and regulation."

Indeed, this paragraph of the NSF OIG regulation sounds rather restrictive until the NSF FOIA office clarified that the "extent permitted by law and regulation" means that *all* documents can be released under FOIA provided they are properly redacted to correspond to the Privacy Act. The "extent permitted by law" refers to redaction and nothing more. The university does not have a legal right to refuse to release the documents.

As an example of how the NSF treats its documents, the full NSF OIG report and NSF letter of reprimand can be retrieved as document A0611054 on the NSF Case Closeout Memoranda search page, https://www.nsf.gov/oig/case-closeout/.[42] It is instructive to go to this webpage and type in the case file number. Or one can select the option to see all the case files. In either case, going to the website to retrieve the document will give the reader a sense of what confidentiality means as practiced by the NSF OIG. There are many other cases on the website. Each of these cases could teach us something about how the process is working. While this information can be extremely useful to individuals involved in specific cases, there is still a general problem in that the scientific community has no idea how many significant cases there are. Although confidentiality is not a barrier to release of documents as University 1 claimed, redaction prevents outsiders from understanding many of the facts of research misconduct cases on file at the NSF. This makes research into the problem much more difficult. While this larger concern looms over the entire research misconduct system, the specifics of each case should be available to the whistleblower and respondents in a way that they can defend themselves. When universities fail to provide the accused with copies of the evidence against them, they are clearly acting in bad faith.

My public records lawsuit ended in mediation in June 2017. This was a good outcome for me as a whistleblower since it provided me a way to salvage my career after years of adverse administrative actions. However, the compromise meant that I did not receive the requested public records. From studying legal cases on file in Wake County I could see that resolution through mediation is rare. The cases are all on file at the Wake County courthouse on an antiquated server based on a 30-year-old technology that predates personal computers. Although all cases that proceed past the preliminary hearing are required to pass through mediation, I could not find an example in the records that involved a lawsuit against University 1 that ended successfully with a mediated settlement. By studying the available cases, I learned that if one does succeed in getting a judgement against the university in Superior Court, the university will appeal all the way to the State Supreme Court. It is a daunting prospect to sue a state university. Since a state university will be represented by lawyers from the State Attorney General's office, the university has the resources to outlast almost any plaintiff. However, a public records lawsuit is in a different category from most as I will discuss further in Chap. 10. Public records law is needed to prevent abuse of confidentiality rules and improper administrative procedures. By regulation, confidentiality rules are supposed to protect the informant and respondent, not the universities. Documents that are withheld by universities in research misconduct cases are actually public records and should be released upon request with appropriate redactions.

7.6 A Confidential Research Misconduct Investigation Can Turn Scientific Fact into Opinion

When graduate students are accused of data manipulation or falsified reporting investigations tend to proceed quickly to a conclusion and the punishment of expulsion from the university is a common punishment [384–386]. When faculty or students discover misrepresentations in collaborative work with faculty, the issue is seldom so simple. It is natural that faculty have the benefit of the doubt when it comes to the way in which data are handled in their laboratories. There is always the possibility that a perceived false statement is supported by evidence or is an honest difference of opinion. Since differences in interpretation are common in science the notion that two faculty members would disagree, but have strong feelings about their interpretation, is readily accepted by almost any observer. The separation of honest error from intentional or reckless falsification is the essence of an exceedingly difficult burden of proof that exists in any investigation of research misconduct. Given how careful scientists are supposed to be, any falsification would seem reckless to most colleagues. That does not change the fact that recklessness or intent must be proven in a legal sense in order to make a finding of research misconduct.

In the hexagon case the respondents repeatedly stated their *opinion* that the palladium-containing reagent was soluble under the solvent conditions that they claimed they used. But solubility data, absorption spectra, mass spectra, thermodynamic data and so on are tabulated values that are not subject to debate. The respondent's claims were contradicted by the data, yet they maintained that they were able to make solutions in their glovebox. They claimed that it did not matter that other laboratories could not make those solutions using the same methods. The other laboratories could not make the solutions, because that was impossible. There arises the ethical question of how to handle the case where the standard of "honest differences of opinion" fails because a given scientific subject cannot be considered a matter of opinion. If a person falsifies data concerning the curvature of the earth and states that one's opinion is that certain incorrect distances must be correct because the earth is flat, should a committee consider there was no intent to falsify because the person's perception was that the earth really is flat? In the hexagon case, the investigation committee ignored the obvious inconsistencies in the data concerning solubility and control experiments and accepted the strenuous objections of the respondents that the entire matter was an "honest difference of opinion". However, the claims made by the respondents were as far from reality as is the claim that there are data supporting the contention that the earth is flat. Since the report is confidential, no outsider will know that the committee made their decision based on unscientific considerations. Any attempt to point this out will be regarded as a punishable offense by General Counsel in the name of the confidentiality regulation.

The respondents went one step further by claiming that a particular glove box was necessary to make the solutions. The respondents told the University 1 committee that the conditions were only correct in the glove box of one of the former University 1 professors and nowhere else. They claimed that there unable to reproduce any of their results at University 1 after 2005 because the special glovebox had been moved to University 2.

> "[One respondent] said in an interview with the Daily Camera that after [the other respondent] left [University 1], taking with him some key pieces of equipment needed for the original experiment, [the first respondent]'s lab was having a hard time replicating [the] original work. But he added, it would have been premature for the team to go to the journals right away with a retraction or correction just because they were having trouble reproducing the work."
>
> "[The second respondent], in a separate interview, agreed. The processes in question were extremely difficult to execute to begin with. One misstep or omission on the part of the researcher could lead to the entire experiment failing."
>
> "[Respondent 1] added that a special glovebox had been used in the original experiment, a necessary piece of equipment he brought with him to [University 2]. That glovebox, a sealed container with attached gloves, allowed researchers to create the perfect conditions needed to complete the experiment as described in the Science article. [according to Respondent 1]"
>
> – Sarah Kuta, Boulder Daily Camera, Feb. 7, 2014

A glove box is a sealed container use to do chemistry in a controlled environment, usually with no oxygen or water present. This is a common piece of equipment used in many laboratories in any chemistry department and various other disciplines as well. The respondents never said what was special about their glove box, because there was nothing special about it. One of the hallmarks of science is that it should be reproducible. If a procedure requires a glove box, then one should specify what conditions are needed, but one glove box should be just as good as any other as long as it is in working order.

I made a demonstration when I was interviewed by the committee. I brought with me sealed vials containing water and containing pure THF with the palladium-containing reagent dissolved. Figure 1.2 shows the appearance of these solutions precisely as used by the respondents. I had prepared the vials in a glovebox identical to the one used by the respondents. I used a gas-tight syringe to extract the purple solution of THF in the palladium-containing reagent and then injected it into water in the correct proportions to make a 10% THF solution. Within seconds, the solution turned cloudy. Within a few minutes, all injected material had been converted into black flakes and precipitated to the bottom. After that, the solution became clear again. Obviously, the committee could see with their own eyes that the palladium-containing reagent did not dissolve in water under conditions claimed by the respondents. Despite this demonstration, the committee wrote in their report that they accepted the respondent's claims that they required the special glovebox to make the 10% THF solution of the palladium-containing reagent. If University 1 General Counsel's incorrect interpretation of confidentiality regulations had prevailed, the report would have been kept secret and no one would ever know that *the special glove box* was part of the reasoning used by the committee in failing to find research misconduct. The point is that pathological science can be accepted under conditions where confidentiality prevents scientific debate [116]. If the reasoning that a special glove box was required had been submitted for outside review it would not have been accepted in any respected journal.

Clearly, the respondents were appealing to the known variability of research results in different laboratories as their defense. But a trained scientist should understand that making a solution should not require a specific glove box. I suggest that the committee members may have accepted such an explanation because they had a hard time believing that anyone who had gone through all the stages of selection to reach the position of a tenured professor would be capable of an intentional falsification. After all, anyone can make a mistake, but why would someone who has a reputation in science ever take the risk of falsifying data? Thus, a faculty member or senior research scientist accused in a research misconduct proceeding will usually have the benefit of the doubt from the research community. It is one of the great failings of the current system that students are seldom, if ever, given such benefit of the doubt. Keeping the reports and deliberations confidential is not in the interest of science or fairness to those involved. One result is that investigation committees can write nonsense that violates the scientific method. However, since it is confidential, most of the time no one would ever know.

Chapter 8
Behind the Façade of Self-Correcting Science

> *Considering its central importance, one might expect replication to be a prominent part of scientific practice. It is not. An important reason for this is that scientists have strong incentives to introduce new ideas but weak incentives to confirm the validity of old ideas. Innovative findings produce rewards of publication, employment, and tenure; replicated findings produce a shrug.*
>
> *– The Reproducibility Project: Psychology, Brian A. Nosek and dozens of authors, Perspectives on Psychological Science, 2012, 7, 657*

The idea that science could have bubbles that are analogous to derivative products in financial markets may seem unlikely given that science can be checked and cross-checked. But is it valid to assume that scientists are checking each other's work with the attention to detail needed to determine whether results are fundamentally sound? In the first place, despite the checks that exist, some scientific results can be sufficiently complicated that it is difficult for anyone outside a field to judge the validity of published results. In the second place, there is no reward, and indeed there are many obstacles, to reproducing published work in subsequent scientific journal articles.

As science has progressed in myriad different directions, specialization has become essential for each individual scientist's success. The consequence of specialization can be so extreme that scientists who are in closely related fields will often feel that they cannot judge the merits of each other's work. Therefore, more than ever before, we must trust that authors accurately represent their procedures and results in their publications. Peer reviewers are supposed to take on the difficult task of determining whether authors have deceived themselves or engaged in questionable reporting because of bias. However, the number of journals has increased to the point that almost every paper that is submitted is eventually published in some journal, as authors move down a list of journals if a paper is rejected in higher impact journals [96]. Once a paper is published it may be difficult for a reader to gauge whether the science is reliable. At least peer reviewers are permitted to ask to

S. Franzen, *University Responsibility for the Adjudication of Research Misconduct*, https://doi.org/10.1007/978-3-030-68063-3_8

see some of the data that are not shown or ask probing questions. But once an article is published it can be difficult to obtain information despite journal policies stating that data should be made available. At the same time, the investments and the promise of new technology mean that the accuracy of the research record is of increasing importance for society. It would be comforting to think that scientists would focus on flawed papers that are in need of correction or retraction by publication of a scientific journal article. The expectation that this occurs is what lies behind the myth of "self-correcting science" [85].

8.1 The Self-Correction Myth Reduces Pressure to Investigate Research Ethics

If one must investigate scientists to hold them accountable for misrepresentations or falsifications, then one must admit that the self-correcting science model has failed. In the article "The Myth of Self-Correcting Science" Sarah Estes uses two high profile cases in the field of psychology as examples where it took decades to detect and correct multiple falsifications [86]. In two other cases of Diederich Stapel and Mark Hauser there have been multiple retractions, more than fifty and ten, respectively [21]. It was whistleblowers rather than publication that exposed these serious cases of scientific fraud. But the self-correction idea is so ensconced in scientists' minds that some will apply it to cases where investigations have been conducted and scientists are found to have falsified data [85]. On the contrary, the fact that ORI and OIGs have had to become involved in policing science is a result of the breakdown of self-regulation [387]. To say that science is self-regulating because frauds are eventually exposed and adjudicated by federal agencies is like saying that neighborhood policing works because murderers are eventually caught by the police and incarcerated. Quite obviously, the goal is to prevent murders in the first place.

The notion of self-correcting science is a convenient concept that provides a relief valve for those who might feel some pressure to speak out about the ethical problems that exist in scientific publications. If self-correcting science fixes the problem, then no one needs to worry about it. The notion that the scientific community will publish and reveal any scientifically incorrect statements (and therefore falsified ones as well) serves as a justification to "move on" and ignore disturbing reports. Scientists are normally anxious to move on because of the high professional price that they would pay for pointing out flawed research. While university administrators have an ethical obligation to investigate an allegation, they too may prefer not to confront the possibility that faculty have published erroneous, misleading or even falsified data if they can possibly avoid it.

How can an administrator justify dismissing an allegation of research misconduct without a proper investigation? Since most of the funds for research come from federal agencies, an inquiry or investigation is subject to the vagaries of the research misconduct regulation in its current form. An administrator has the right to decide that an allegation is frivolous, which would be grounds to reject it without further inquiry. The financial pressures on an administrator can also provide a reason for downplaying a potential ethical issue. If one can convince oneself that the issue is a scientific dispute that *should be hashed out in the journals*, then the moral obligation to conduct an inquiry is removed. However, in reality the only time when a research group will focus on a correction of published errors is when it becomes absolutely essential because it conflicts with a core finding of research. Usually this correction is needed to ensure that funding is not affected because of some misunderstanding of the veracity of key research. On the other hand, the idea that there is some general tendency for errors to be corrected by other scientists is a way of dismissing an uncomfortable topic and nothing more.

The matter is quite a bit more serious when dubious publications support patented technology. In a private company one would have to be concerned about potential allegations of fraud in instances where the claims of the technology turn out not to be supported. In such cases the published correction may have important implications for the validity of a technology. Imagine a case where technology has been patented and then a correction appears in the scientific literature. If private funds have been invested in the technology, then the scientists must answer to investors. The investors will want to know if the technology has been invalidated by a new report that has conflicting evidence. Discovering discrepancies in published work is difficult for any scientist and there is a case to be made for not rushing to judgement. If the correction stands the test of time, the scientists and investors may be forced to accept that published data could be wrong or even falsified. In such cases, the ethical imperative is clear. The problem that arises in real life is that there are ways to deflect the criticism and to try to save face even when doing so violates ethical norms. The self-correction concept is based on the faith that others will recognize when they need to revise their results or even retract a publication. When the stakes are high enough people may seek other ways to diffuse the situation.

There are only a few studies of self-correcting science, but in one such study published in Nature, Dr. David Allison and his colleagues searched the literature on obesity, nutrition and energetics for errors [388]. They identified three common errors in the studies in their field. 1.) mistaken design of randomized trials 2.) miscalculation in the analysis 3.) inappropriate baseline comparisons. Errors 2.) and 3.) are essentially a common refrain of poor statistical analysis corresponding to poor estimation of the variance and incorrect definition of the control group, respectively. In Allison's study, only one author was receptive to the results of the analysis and agreed to retract the flawed publication. Twenty-five others refused to correct their mistakes. Allison's study directly challenged the notion that the literature is self-correcting. Along these lines, independent statistical analyses have played a major role in the detection of falsification and fabrication of clinical trials [326, 327]. While the increasing awareness of statistical analyses and the rise of post-publication peer review have had an effect, it is individual scientists who are leading these efforts. Universities are, overall, dragging their feet and acting as though the problem is not serious enough to warrant greater attention.

8.2 The Failure of Self-Correction in The Hexagon Case

The ideas in the 2004 *Science* paper and related technologies called "RNA-mediated materials synthesis" were marketed to investors as the research misconduct investigation continued. The patent application (PCT/US2004/029480) [31] was used from 2004 to 2007, and other patent applications were filed subsequently when the respondents moved to University 2 with the intent of starting a research center based on the idea. Our attempts to publish a correction in the scientific literature during this period were greeted with both public and private attempts to derail the scientific "debate". While the respondents were sending the threatening letters to editors and to University 1 in private, the respondents pointed out numerous times in public that there were no others joining our research group in publishing corrections to their work. They did not mention that, by this time University 1 had formally, but confidentially, requested that the respondents correct the record in a letter written following the conclusion of the university research misconduct investigation in 2008. Under significant pressure the respondents began to publish "corrections" in 2009, but none of these actually corrected the research record. No editor, reviewer or university administrator appeared to notice glaring contradictions that invalidated each of the corrections. In 2016 it was revealed that the agents at the NSF OIG were keeping track of these discrepancies and contradictions, which were tallied in the final OIG report [42]. It is ironic to speak of self-correcting science when the only institution that was monitoring the corrections was a law enforcement agency.

The scientific community will never know the actual number of other research groups that attempted to work on the method report in the 2004 *Science* paper and simply gave up because they could not reproduce it. I know of two research groups that wasted significant time and resources trying to replicate what was impossible, because it had not been reported accurately. Once sufficient time has been spent to be sure that all efforts have been made to replicate a given method, most scientists give up when they cannot reproduce a result. At the end of such a frustrating and wasteful experience, it is essentially impossible to publish anything from the effort. Scientists will prefer to move on to do something productive instead. They do not even want to discuss the matter. However, two professors were kind enough to tell me privately of their experience and the bitterness they felt having believed the report in Science that led to waste of significant time and money. No one wants to get involved in such sordid matters so I am the only person who knows about these efforts. The papers that my research group published and that I published on my own were a response to the fact that my reputation as a scientist was under attack because of the role I had played in reporting the misrepresentations made by the respondents. Our publications on the hexagonal particles required a great deal of effort. Every scientist knows that journal articles must be new contributions, not a rehash of other's failed experiments. The publication record described in the case of the hexagonal particles reveals the fallacy of relying on publication alone to correct an error, particularly when it involves a breach of ethics. Errors can be entrenched for many reasons, but if there is a financial motivation for protecting a particular result in the scientific literature, it will take more than a publication to correct the research record.

The role of the individuals seems forgotten in the "self-correcting science" myth. Consider the reality faced by the scientist who wants to correct the record. Publishing a paper that corrects errors in instrumental analysis or omissions of data must meet criteria for independent science that show an advance in the field. One cannot publish a paper that simply points out errors and inconsistencies however serious they may be. The paper will be peer-reviewed by reviewers who will ask whether the contribution is a novel contribution to science. They will ask what significant advance has been made. The editor will insist that a scientist who publishes to correct the record will be held to the same standard as anyone else. They must have *new* data, *original* ideas and *innovative* methods. How does one couch the message of correction in language that heralds a new discovery or insight? If the author says directly what the goal of the paper is to correct an error, then by definition the work will be regarded as derivative and rejected. One must do more, much more with an idea to warrant publication in a journal that anyone will pay attention to. But in the current climate where ideas are patented and scientists are using those ideas to try to start companies, if one succeeds in publishing an article that calls someone's work into question then they will probably respond forcefully to protect their interests.

8.3 The Changing Attitudes Toward Data Sharing

Publicly funded research is supposed to be transparent and the results are intended to be shared. Any scientist should be able to repeat the process and therefore all scientists have a right to examine the data and analysis. Thus, even if the work had not been jointly funded, we would have had a right to analyze data and present our findings and opinion about what those data indicated. Unfortunately, the idea that data can be owned seems to be gaining traction in universities. Patent laws have changed in recent years and there are now many new ways that outsiders can threaten research done in universities. I am not criticizing patenting *per se*, but rather the indiscriminate use of perceived patent position to squelch innovation or checking of results by other scientists.

How can someone credibly claim that a former collaborator had no right to access data or publish results? Of course, it would be unethical to enter someone's lab and take their samples and or their data, which is what the respondents were implying I had done. But on the contrary, in this case we had obtained the funding jointly and studied the samples in collaboration. Even if this were not the case, by convention and according the publication policies of most journals (including the journal *Science*) anyone should have the right to request the data or even samples for further analysis. Although I asked to see the original data by way of the editors, the editors of *Science* wrote to me that they could not force the authors to provide the data. The data we requested complemented that vast amount of data we already had obtained as members of a collaborative team. There should have been no question that we all had the right to access the data in our lab and paid for by our grant, but this was questioned and even became the subject of a research misconduct investigation initiated by University 1.

The notion that falsifications can be protected as intellectual property is perhaps the most disturbing aspect of the hexagon case. Despite the claims of intellectual property, the patent on the work in the *Science* publication [31] was abandoned before our 2007 *JACS* paper was published [24]. Of course, falsified data in a patent application is fraud and there are legal consequences if that is proven in court [389]. Thus, it is perhaps not surprising that while maintaining publicly that the result was not falsified, the patent was permitted to lapse in 2007, before the research misconduct investigation at University 1 was completed. Nonetheless, even several years after the patent had lapsed, the existence of the patent served as a pretext for claiming that our publications were intellectual property violations. When intellectual property is used as a legal threat, it is difficult for a university or editor to know whether that threat has any validity, but it will add to the suspicions that a person who is trying to correct the research record has ulterior motives. Indeed, "self-correction" is an outdated concept in an era when patenting inventions and ownership of data as intellectual property are on the rise.

Chapter 9
The Origin of the Modern Research Misconduct System

> *"During the past two decades, the process of professional self-regulation in the scientific community has been seriously questioned by the lay public in general and Congress in particular. Congressional concern about failures in self-regulation are manifested in their establishment of the National Science Foundation's Office of Inspector General and the Public Health Service's National Institutes of Health's (NIH) Office of Scientific Integrity to investigate and sanction scientific wrongdoing."*
> *Braxton and Baird "Preparation for Professional Self-Regulation"*
>
> *Science and Engineering Ethics, 2001, 7, 593*

In response to the ethical crises in the 1980s, an act of Congress mandated that a research misconduct system should be implemented. Throughout the 1990s meetings were held by panels of scientists and other stakeholders to draw up appropriate language for research misconduct regulations. The final report issued in 1999 by the Office of Science and Technology Policy (OSTP) requested each federal agency to draft its own research misconduct regulations. By 2005 each agency had regulations in place using common language that recognized only "falsification, fabrication or plagiarism" as grounds for an allegation using the NSF OIG regulation (45 CFR § 689.1 *et seq.*) as the model for federal regulation. The input from scientists and universities led to a system in which the universities have the primary responsibility for investigation, but agencies have oversight authority. The failure to recognize the inherent conflict of interest in university conduct of investigations is more than a minor oversight in the policy now that universities have become centers of commercialization. Although originally intended to protect informants and respondents, confidentiality has been used as a shield to protect the university from bad publicity. University general counsels have often interpreted confidentiality rules to the detriment of science and those scientists involved in an investigation. Thanks to the confidentiality clauses in the federal regulations we are being carefully protected from information that would document how scientific fraud has become a pervasive

S. Franzen, *University Responsibility for the Adjudication of Research Misconduct*, https://doi.org/10.1007/978-3-030-68063-3_9

problem in our universities. The system has the effect of protecting senior research scientists unless the case involves sexual misconduct or financial impropriety. The Taleyarkhan bubble fusion case at Purdue University is the only publicly known case where a university has been compelled to revisit a conclusion under federal scrutiny [373].

In common with many sectors of society and the economy, scientists have resisted regulation by the federal government. Despite the testimony of Dr. Handler in the fateful Congressional Hearings of 1981, Congress decided that scientists had failed to make a compelling case that they should be exempt from ethics regulations. The long process of building a consensus as to what should constitute research misconduct policy began in 1985. It took 20 years for the regulations to be written in final form in the OIGs of various federal agencies that funding science, such as USDA, EPA, DoD, DoE, NSF and in the ORI of the NIH. Shortly after the law passed Congress in 1985 mandating these regulations, a case of research ethics became front-page news and stayed in the spotlight for a decade. The case had a strong impact of the opinion of many scientists and led to their input to try to minimize the extent of the regulations. In part because of the extreme publicity of the Baltimore case, all stakeholders agreed that publicity was disastrous and thus confidentiality became an important aspect of the new regulation.

9.1 The Baltimore Case

The Baltimore case arose from allegations concerning an article published in 1986 in the journal *Cell* [390]. During the ensuing decade the case became a defining event that shaped the consciousness of the scientific world with regard to research misconduct and its significance [10, 332, 391]. The primary data reported in the laboratory notebooks of Imanishi-Kari at Tufts University were questioned by a post-doc named Dr. Margot O'Toole. One day in the spring of 1985, O'Toole was looking through laboratory records for a particular genetically modified mouse when she discovered seventeen pages of laboratory notes. Since she had studied the recommended paper by Weaver et al. on a daily basis for previous 11 months and was having difficulty reproducing certain aspects she immediately understood that the notes showed that the paper had misrepresented the protocols actually used in the laboratory [390]. One of the co-authors, Nobel-prize winner, Dr. David Baltimore who was the Director and founder of the Whitehead Institute at MIT got involved in the case at an early stage to defend Imanishi-Kari. Dr. Baltimore was never himself a respondent, but because of his strong involvement and notoriety the case became commonly known as the "Baltimore case" [391, 393].

Despite an initial recognition of significant discrepancies in the record by administrators at Tufts University, the administrators who met with O'Toole at Tufts and Massachusetts Institute of Technology dismissed O'Toole's allegations of data misrepresentation. Although these meetings were later called "investigations" they were not even inquiries by the current standard for response to an allegation of research misconduct [7]. An undergraduate, Charles Maplethorpe, who had worked in Imanishi-Kari laboratory notified Drs. Feder and Stewart at the NIH, who became involved when they learned the facts of the case. The Baltimore case arose very shortly after the formation of ORI, which had not yet been relocated from HHS to the NIH. The ORI became responsible for conduct of the subsequent investigation, but the public nature of the investigation and involvement of many other branches of government was unusual by today's standard. The authenticity of notebook pages was investigated by the Secret Service. Later the handling of the allegations became the subject of Congressional hearings. Baltimore used his position to deflect any criticism and made the fundamental mistake of not first studying the notebooks and revisiting the science. Perhaps he did not realize that Dr. O'Toole found an inconsistency between the laboratory notebooks and claims in the article, which appeared to be incorrect in the sense that they could not be reproduced in the laboratory.

The scientific observation in the *Cell* paper was that the introduction of foreign genes for an antibody into a mouse stimulated the production of related antibodies [393]. The central claim of the paper was a validation of Jerne's network theory of immune system regulation. But that theory was not substantiated by the paper. Baltimore later denied that this was the point of the paper. But other scientific observers have stated that it was clearly the intended point of the paper [7]. Based on his behavior it appears that Baltimore was not arguing a scientific point, but rather a point of personal pride. He was adamant that he was not involved in a publication that was wrong. As happened in the hexagon case, the scientific research and publication proceeded even as the case was being deliberated. In 1993 Imanishi-Kari published two articles that she claimed redeemed her original idea [394]. In 1993, Dr. Baltimore *unretracted* the paper and restored it to the journal *Cell* [390]. Because of the press coverage the case became highly controversial with the one of the most public moments consisting of Baltimore's testimony before the U.S. Congress. Dr. Baltimore was convinced that the government had no business getting involved in scientific "disputes". Baltimore organized a letter writing campaign to portray the hearing as a witch hunt and an attack on science. Congressman Dingell from Michigan was equally convinced that the evidence had not been taken seriously and that O'Toole had received poor treatment by the universities and the scientific community. In 1994, the ORI found Imanishi-Kari guilty of research misconduct. But then, in 1996, Baltimore and Imanishi-Kari appealed the decision through the newly established ORI appeals process. Unlike the investigation, which was conducted at ORI by a committee of scientists, the appeal was a strictly procedural venue with three judges and a lawyer for the prosecution and defense. Judson in his book "The Great Betrayal" describes the travesty of the appeals process. The matter was adjudicated by a three-person panel, two lawyers and an emeritus professor of biology, who ruled only on whether the ORI had followed protocol [7]. Judson, who was present at the appeal, writes about the judges' evident bias. The judges were willing to dismiss evidence for almost any reason. For example, a mistake in the start date of the Secret Service work provided by one witness resulted in exclusion of *all* the forensic evidence of tampering with the notebooks. In the end, Baltimore succeeded in the appeal. The conviction by the ORI was thrown out based on procedural considerations. The ORI process in those early days was conducted strictly on scientific grounds. The New York Times dubbed it the "case that evaporated" [395], but it would be more apt to call it the "case that exploded". By the time of the final appeal, the Imanishi-Kari paper had already been retracted, reinstated, with publicly known involvement by the Secret Service and NIH, the ORI had made a finding of misconduct and there had been highly public drama in four Congressional hearings [396]. No doubt a Congressional hearing is not the proper venue to discuss scientific fraud, but the response by Baltimore and many of his supporters was to deny that there was a problem at all. Once again, the leaders of the scientific community denied that scientific ethics is an issue worthy of discussion.

When a case goes before an appeals court judge the entire scientific basis of the case is moot. All that matters is whether the protocols in the regulation were followed to the letter. The regulations are so vague that it is easy to make a case that procedures were not followed. The Baltimore case is an example where strong evidence of misrepresentation of data was discounted on appeal. The duration of a case itself can become a reason to throw the verdict out upon appeal. For example, both the Baltimore case and the hexagon case took roughly a decade to adjudicate. That was not because of a lack of evidence, but because of bureaucratic inefficiency. Even though the agencies are underfunded and a large reason for the delay is the crushing case load, that very inefficiency could itself become a reason to invalidate the case on appeal. After all, the respondents may argue that it has been grossly unfair to them that their careers were put on hold for the duration of the case. The evidence may be discounted because of errors in bookkeeping as happened in the Baltimore case. When such reasoning is used to dismiss the case, all that any outsider will know about the case is that the respondents were "exonerated". The perverse outcome leads us to question how research misconduct should be defined. Author Horace Judson quotes Baltimore's close colleague and friend Howard Temin in an interview on March 16, 1993 [7]. "David's misconduct was – When an experiment is challenged no matter who it is challenged by, it's your responsibility to check. That is an ironclad rule of science that, when you publish something you are responsible for it. And one of the great strengths of American science, as opposed to Russian and German and Japanese science, is that even the most senior professor if challenged by the lowliest technician or graduate student, is required to treat them seriously and to consider their criticisms. It is one of the most fundamental aspects of science in America."

9.2 Consequences of the Baltimore Case

The Baltimore-Imanishi-Kari case, perhaps more than any other, defined the research-misconduct battle lines within the scientific community, university administration and federal agencies. There is still no general agreement about the lessons to be learned from the case [7, 10, 397]. Baltimore maintained that scientists can and should regulate themselves and that the government has no role to play in adjudicating scientific ethics. However, Congressman John Dingell countered that the scientists had bullied and persecuted O'Toole, the whistleblower in the case, rather than attempting to understand the significant evidence of falsification. University administrators showed their reluctance to take any action even when presented with overwhelming evidence. The federal government ultimately handled the matter by finding research misconduct, although the reversal in the finding made by the ORI in a final appeals hearing ensured that no one in the scientific community would be satisfied with the outcome of 10-year ordeal.

It is a remarkable coincidence that Dr. Baltimore suffered an even greater scandal 8 years later when Dr. Luk van Parijs, a former post-doc of Dr. Baltimore's while at CalTech was found to have committed research misconduct involving complete fabrication of research results in several publications [398]. Van Parijs was found guilty of research misconduct and because he had falsified data on a federal grant application he was also found guilty of grant fraud. Once this finding was published, it was revealed that van Parijs' lack of ethics had been reported 15 years prior to his sentencing. Baltimore could have learned about it if he had vetted his post-doc before hiring him at CalTech since others had expressed concerns at an early stage van Parijs' career [399]. It is almost as though scientists have put on blinders that prevent them from seeing any evidence of misconduct even when it is right before their eyes. Baltimore pleaded for clemency at the sentencing so that van Parijs was spared jail time [400]. The whistleblowers in the case were students at MIT where van Parijs had begun his academic career. Many of them wasted years because of van Parijs' dishonesty. Margot O'Toole, Charles Maplethorpe and many others were affected by the unethical actions of people who worked under David Baltimore's umbrella. These cases show that the hubris scientists and misconduct tolerated by the scientific community can result in great harm to others. Yet, the harm caused by research misconduct retaliation is not considered in the regulation.

The example of Margot O'Toole shows that a post-doc or graduate student will have a difficult time getting a fair hearing, particularly if their allegation concerns an established scientist. For a post-doc to stand up to a Nobel laureate is the most extreme example of a difference in status that has been discussed in the literature on research misconduct. Congressman Dingell wrote an editorial stating that it was the lack of openness on the part of the scientists that led to the escalation of the Baltimore case into a major case of research misconduct [332]. The investigation incited fears among scientists that the research misconduct process would be politicized. In a response to Baltimore's statement that scientists should be left to police themselves, Congressman Dingell wrote that scientists had abdicated their responsibility [332]. The controversy underscores the psychology of research misconduct investigations to this day. If the respondents are established scientists, they will be able to garner tremendous sympathy from colleagues who trust that they would not falsify data. The corollary to this sympathy for the respondent is antipathy towards the whistleblower. If it is difficult to conceive that the senior scientists manipulated or omitted data, then it is possible to ascribe pernicious motives or incompetence to the informant.

As a result of the public nature of the conduct of the Baltimore case, it was agreed in both government and scientific circles that a research misconduct policy should be set forth that would permit a reasonable and deliberate resolution of allegations of falsification. While leaders in science had attempted to prevent the formation of a formal body to investigate research misconduct the Baltimore case made it clear that there was dysfunction in how scientists themselves had managed these matters. In the early 1990s possible remedies were studied in a task force at the National Academy of Science, which made recommendations to the OSTP in the White House. Ideally, the policy would give scientists the possibility to adjudicate any potential misconduct. On the other hand, the responsibility of the federal agencies as stewards of public funds became the driving force for the policy. "Research misconduct" as defined by the OIGs became the focus of the federal regulations that were drafted in response to the OSTP directives [110]. It was assumed that delegating the actual inquiry and investigation to the universities would accommodate the need for scientists to have input into the process of adjudication and to avoid heavy-handed government interference. While scientists are the most qualified to judge whether data have been manipulated, misrepresented or falsified, the regulation converts the major deliberation into a legal question by the emphasis on "intent" or "recklessness" as necessary conditions for a finding of research misconduct. It also turns the process into an administrative-legalistic one within the university with an unfortunate consequence that scientific considerations take a back seat. Furthermore, federal OIGs, or the ORI, are interested only in determining whether the funds were properly spent. They have no mandate to examine how inaccurate information in the journals is managed by editors. The taxpayer and scientific community are two separate stakeholders and the regulation does not adequately separate their interests. Of course, the federal agency needs to be involved when there are cases where funds have been misspent because of falsified research but scientists need to be more involved in the decisions regarding the correction of the scientific literature because distortions of the research record can cause great harm. In the final form of the regulation, the scientists gave all the authority for investigation and the recommended adjudication to university administrations.

9.3 The Path to Federal Oversight of Research Ethics

The path to the present regulation started in 1985 when Congress passed the Health Research Extension Act, which required the HHS to establish a regulation for fraud in science [401]. In 1989 the HHS created Office of Scientific Integrity (OSI) with the NIH. The OSI merged with a similar office in the department of Health and Human Services (HHS) in 1992 to become the ORI. As the pioneer agency in the conduct of research misconduct cases, the ORI had numerous issues to overcome including training and consistency of procedures, as well as meeting the regulatory timeline for cases [379]. In 1999 the decision was made to extend the concept of an integrity office to each government agency that funds scientific research. The current regulation is based on the OSTP publication in the Federal Register (October 14, 1999 pp. 55722–55725) with subsequent revisions made in 2005. A parallel trajectory within the NSF started in 1987 with the proposal of a new set of rules, which became the research misconduct regulation in 45 CFR § 689.1 *et seq.* Although the ultimate power to adjudicate research misconduct rests with the federal agencies, NSF regulation 45 CFR § 689.4 states that universities "have the primary responsibility to maintain the research record". The HHS regulation is not specific about the role of institutions should play in maintaining ethical standards. In § 93.319 states only that "an institution may find conduct to be actionable under its standards even if the action does not meet this part's definition of research misconduct." [402] The relationship between the research institution and federal agencies is stated most clearly in the DoE regulation 10 CFR Parts § 600 and § 733 where it states that "The core principle of the Federal Policy is that, while research organizations have the primary responsibility for the inquiry, investigation, and adjudication of allegations of research misconduct, federal agencies have ultimate oversight authority for the research they fund or support." [403]

Although it has a similar definition of misconduct, the ORI has more limited powers than the OIGs and it is not capable of conducting an investigation. Among the various OIGs, only the NSF OIG conducts independent research misconduct investigations. The NSF OIG will undertake an investigation only if it finds a shortcoming in the university process. For this reason, the NSF OIG can be regarded as the model organization for the current system. Both ORI and NSF OIG will follow the regulation and are responsive when presented with allegations, which means that they refer investigations to universities or other "awardee institutions". However, by regulation neither the ORI nor the NSF OIG have any oversight role in an actual investigation, unless a university overtly refuses to take any action or fails to take complete and appropriate action. As long as the university answers a referral letter and states that they have conducted an inquiry the ORI or NSF OIG will normally accept that the university has discharged its obligation it in good faith.

9.4 The Handling of Evidence in Research Misconduct Cases

The standards for allegations, evidence and testimony in research misconduct processes are quite different from those in a court of law. The procedures are much less fair to the accused (respondent) since every aspect of the investigation and deliberations is kept confidential. The accused has no right to know their accuser, to know what the evidence is or what the testimony is against them. Some universities may share that information with the respondent, but it is not required. On the other hand, these procedures are also unfair to the whistleblower since the testimony by the informant is not required and may be ignored. The committee is not required to hear the whistleblower's statement or ask for any clarifications regarding the allegations or evidence. Since the whistleblower will likely suffer negative consequences in the event the committee fails to find research misconduct, the presentation of all the evidence is crucial to the conduct of the case. Failure to properly conduct a case will have negative consequences for the whistleblower.

The process has certain attributes of a secret court. Federal regulations merely state that any determination should be made based on a preponderance of the evidence without specifying any rules of evidence. For example, there is no rule that says that evidence supplied by the whistleblower must be forwarded to the committee. The RIO may define what is considered evidence and how to notify the committee of its existence. Frequently, the RIO is not a scientist and therefore lacks the expertise to evaluate the relevance of evidence. It is assumed that the committee will ask the witnesses questions based on the evidence. However, without a prosecutor or defending attorney how does the committee know where to look in the evidence to piece together the scientific thread in a complex case? The process resembles the blind leading the blind.

The ORI regulation states that evidence should be sequestered at the inquiry stage and then examined by the inquiry or investigation committee [402]. However, even the ORI regulation does not provide the same kinds of rules of evidence that exist in a court of law. Most other agency regulations make no mention of evidence. The agencies do have their own internal regulations although members of the public have no way to know what these are. Agency referral letters may contain instructions for the university to sequester evidence. However, it is a remarkable state of affairs when a citizen cannot even know what the protocols are for government agency investigations. I once asked an OIG agent if I could see the agency regulations at NSF OIG. The agent told me that agency regulations are "confidential". If one is concerned that the university process is flawed the federal agency that has jurisdiction does have authority to investigate. Nonetheless, no one has the right to know how the investigation will be conducted or how long it will take. Of course, the agents at the responsible OIG are supposed to read the university report after an inquiry is completed. However, as has been shown in a court case, federal agents do not feel that they are required to read the report to come to a conclusion [404]. Even if they do take their charge seriously and read the report, how would they know if evidence was ignored simply by reading a university report?

9.5 The Constitution of an Inquiry or Investigation Committee

Research misconduct regulations in the various agencies provide no guidance as to how to constitute an investigative committee. Consequently, the procedure for defining a committee varies from institution to institution. For example, the committee at University 1 is convened by invitation of the RIO. This procedure gives the RIO the flexibility to call on experts in various fields appropriate to the case to serve on the committee. However, this is no easy task for the RIO since there is no incentive for faculty to serve and a large potential downside in terms of wasted time and entanglement in a controversy that can interfere with one's career and even one's peace of mind. The notion that tenured research faculty will feel a sense of responsibility to serve misses the point. The very proximity one has by virtue of being a faculty member in the same institution as the informant and respondent is uncomfortable to the point of overwhelming any sense of civic duty. University 2, on the other hand, has a Standing Committee consisting of a wide range of faculty from across the university. The advantage of the Standing Committee is that one does not need to spend time to assemble a committee for each investigation. The Standing Committee is larger than an ad hoc assembled committee. For example, an Inquiry and Investigation Committee at University 1 must consist of at least three and five people, respectively, while the Standing Committee at University 2 consists of eleven people. However, each of these ways of constituting a committee has potential pitfalls.

If an ad hoc investigative committee is constituted to examine allegations of research misconduct against a faculty member, one is asking the committee to pass judgement on a peer. If the subject of the inquiry or investigation is a full professor then one should, in practice, only ask full professors. Otherwise, one is asking for peers of lesser rank to make the judgement on a superior rank within the system. This requirement severely restricts the number of potential committee members. Of course, the greatest difficulty can reside in the evaluation of the scientific merits of the allegations. Scientific experts must be involved in an examination of the evidence for the committee finding to be credible. This further restricts the number of qualified committee members and increases the risk that there is a perceived or real conflict of interest. When two scientists work in the same or related fields, there is always a potential that they have come into contact professionally either as competitors or collaborators. Almost all these individuals know each other. The reader must recall that the committee members will be asked whether they believe that their colleagues have falsified data with "intent" or "recklessness". Asking full professors to examine the motives of their colleagues puts them into a most difficult position. Do we really expect senior faculty to participate voluntarily in an unpaid assignment of this type to judge their own colleagues? Often the faculty members know that they will be in a lose-lose situation since either the informant or the respondent will suffer because of their conclusion and in some cases, both are colleagues. Of course, most faculty will declare that they cannot participate due to a conflict of interest. They are correct to do so if they know the respondent or whistleblower as a colleague. They would surely have a difficult job remaining objective about the case. But who then can be summoned to serve? If the committee completely lacks an understanding of the field, it can be difficult to assess the merits of the case even when there is no conflict of interest. When conflicts exist, there can be no expectation of an impartial expert evaluation of allegations based on evidence.

9.6 The Handling of Allegations in Research Misconduct Cases

Given that a research misconduct committee works in isolation, the writing of the allegations of research misconduct is the single most important step in the investigative process. Nonetheless, a university ethics or sponsored programs website will typically offer relatively little or no guidance regarding the crafting of allegations. The suggestions that exist for formulation of the allegations, such as the vague guidance on the ORI website, do not include details such as how to include the evidence and how to precisely document the observations of misconduct. A consistent and fair process should quickly determine whether there is any merit to the allegations. However, once the allegations are submitted, the whistleblower will likely have little or no further input into the process. Poorly written allegations can lead to misunderstanding by the committee and it is likely that there will be no attempt to seek clarification.

If an allegation is filed at a university, the RIO may deem the allegation to be frivolous, and therefore the university would not even be required to notify the appropriate federal authority. An allegation may have substance, but it may be deemed frivolous because it is poorly written or because of institutional conflicts of interest. Given the lack of experience of most whistleblowers with the art of writing research misconduct allegations, it is possible to write allegations that will not be taken seriously, even when there is strong evidence. Some estimates suggest more than 70% of the allegations filed at universities, medical schools and other institutions are not pursued [405]. If the allegation is filed at the federal agency OIG there is less chance for this outcome since the federal authority will often notify the university with a referral letter, which requires an inquiry at a minimum. Breen discusses the reluctance of whistleblowers to come forward in a frank article that asks if the research misconduct regulation needs to be reconsidered [406]. The article concerns the Australian experience, but like many outside the U.S. the reference point for comparison is the research misconduct regulation of the U.S. agency OIGs. The international experience is still quite relevant since most countries (particularly in the English-speaking world) have patterned their research misconduct policy on the American version.

One would expect a good faith whistleblower to feel conflicted as soon as it is necessary to write an allegation. It is difficult to make an accusation, particularly against colleagues, even in the face of overwhelming evidence. If one is convinced that data have been manipulated, omitted, or otherwise misrepresented, every detail of the treatment of data should be spelled out with evidence items numbered to avoid any ambiguity in interpretation. The allegations should document how the behavior is a deviation from the norms of accepted practice. The informant should not assume that anyone will look into the record and put the pieces together. Rather the informant needs to write a standalone case that explicitly ties the evidence to each allegation and explains evidence for intent or recklessness. But without having the benefit of understanding the trajectory of a case, a knowledge of the regulation and the rules of evidence, it is unclear how most scientists would know to write proper allegations without guidance. If a scientist or perhaps a research group as a whole has done something that involves wholesale fabrication, it can be very difficult for the whistleblower to know the details *a priori*. For example, in the Taleyarkhan case, it was alleged that Dr. Taleyarkhan had published a paper without his own name to try to prove that there was independent confirmation of his result. Because of confidentiality we do not know how an authorship issue turned into a case that lasted many years and ultimately prompted the House Subcommittee on Investigations and Oversight to request that funding sponsor revisit the conclusion [373, 374, 407]. The system is simply not working well when something as simple as an authorship question becomes a task that requires years of deliberation and involvement by government agencies to ensure compliance.

9.7 Punishment in the Research Misconduct System

The nature of the punishment in research misconduct cases can lead to a charged atmosphere in the scientific community. The punishment set forth in the regulation is 3 years of debarment and certifications, which means that a scientist's publications must be monitored for 3 years to ensure compliance. There is no requirement that the university act to correct or retract publications found to contain falsified or fabricated data. While it was most likely seen as common sense that universities would act to correct the record, this aspect of the regulation has not been a success. Retraction is a serious matter and universities fear loss of prestige or lawsuits from affected faculty. Despite receiving no mention in any of the federal regulations, retraction is actually the most appropriate punishment. Most scientists are concerned primarily with the integrity of the research record. For a scientist, the punishment of a forced retraction would be severe enough to have a profound deterring effect. By turning the matter primarily into a financial punishment, the focus is on a more severe kind of wrongdoing in the eyes of the public and of fellow scientists as well. This focus occurs at the expense of concern for the research record.

Because of the nature of punishment, the research misconduct process is seen as outside interference in matters of science. Many scientists imagine that research misconduct would spell the end of a career. In an article in Nature that examined the trajectory careers following research misconduct, half of scientists continued to be active [408]. Because of the potential severity of the repercussions, the actions of the whistleblower may be seen as irresponsible unless the evidence is clear and overwhelming. But then how would others know whether the evidence is strong since considered a violation of confidentiality to inform anyone of the substance of the allegations. In the normal conduct of science, one would always show the evidence for any claim.

There are observers today who clamor for more severe punishment for research misconduct. Perhaps this too is the logical outcome of the Bayh-Dole act. When professors can own intellectual property, which is paid for by taxpayers, there is a responsibility that comes with the new reward structure. If research is conducted purely for the pursuit of knowledge, it is hard to imagine the motivation for punishment in a legal sense. However, the social contract of the university system is changing. It is within the realm of possibility that research misconduct could be treated as a crime in the future. Of course, it is already a crime to misuse taxpayer funds, but that includes such things as paying for vacations or cars using federal grant money, which is fraud, and not research misconduct. This distinction is clear in the NSF Office of Inspector General where there are two separate divisions for the different types of investigations. Rather than implementing criminal penalties for individuals, perhaps the federal agencies should hold institutions accountable. Universities would be more responsive if the agencies began to ask them to return the funding that supported falsified research [409]. When one considers this possibility it is obvious that the universities cannot be placed in the position of investigating their own potential malfeasance.

9.8 Ambiguity in Legal Definitions in Research Misconduct Regulation

The role of lawyers in research misconduct inquiries and investigations is not defined in the federal regulation. Technically speaking, lawyers retained by the respondents (and more rarely by the informant) are considered to be spectators who can ensure that the rules are followed. But the reality is that legal challenges made in briefs sent to the RIO and university general counsel can significantly delay a research misconduct investigation. An unscrupulous respondent may use a lawyer to intimidate the committee and the administration. A lawyer will look for technicalities to have the case thrown out and will remind those present that there is always the possibility of a lawsuit if the case turns out badly for the respondent. The regulation opens the door to this since 45 CFR § 689.10 permits an appeal in a court of law if there is a finding of research misconduct.

The involvement of lawyers exacerbates an already strained system. Lawyers can apply pressure at every step to slow the process or to threaten to derail it. University participants in the process, who are usually not familiar with the law may be easily intimidated. The RIO is usually not a lawyer and will need to refer any brief received from either side's legal team to the university General Counsel. The research misconduct committee, which is seldom compensated for extra effort, may also be concerned about being drawn into a legal battle after their decision. The fact that the university has the responsibility to represent the members of the committee in the event of a lawsuit is little consolation for the time lost and stress of being sued. Yet, despite the legal threats hanging over the heads of the committee members, the format for committee meetings envisioned by federal regulation is not a legal proceeding, but an interview by an academic committee. Working against that format, university General Counsel will have no stake in the scientific issues. General

Counsel will typically look for exit strategies if there is a significant potential for a lawsuit. Avoidance of a lawsuit leads to decisions that favor the side with legal representation (e.g. limiting access to evidence, limiting questions to witnesses etc.) Legal considerations of this type are detrimental to scientific deliberation. Since the science is the subject of the investigation by definition, the regulation should strictly define the role of legal considerations. The regulation not only fails to do this, but presents such ambiguity that university lawyers can interpret the regulations in accord with their self-interest.

The absence of any consideration of the harm done by research misconduct is a profound weakness in the regulation. Even errors can have a major negative impact on the overall development of science, but falsification and fabrication have a much greater potential to negatively impact other scientists. To illustrate this I choose an example that is clearly not misconduct, but an egregious error. A completely distorted structure of the ABC transporter protein was caused by a sign error in an analysis program for the diffraction data. This leads to the wrong overall topology and incorrect interpretation of biology. This distorted structure of an important membrane protein was published in five top journals only to be discovered to be incorrect 5 years later. The damage done to the community was enormous. Incorrect conclusions were propagated in many related fields that relied on the structure. Certain scientists failed to get funding because their laboratory results were apparently inconsistent with the published structure. Only because of the persistence of a Swiss competitor Kaspar Lochar was the error discovered and the record set straight [410]. Based on this case of a mistake, one can begin to imagine the consequences when a fraud is published and remains in print for years. Misleading others can be disastrous for their careers. Graduate students are beholden to their advisor for approval of the PhD (or MS) and for recommendations. They are in a vulnerable position and can be manipulated by an unscrupulous professor. The professor-student relationship is largely based on the honor system and there are few mechanisms that enforce the expectation that professors will act ethically and keep their students' interests foremost in their minds. A great deal of scientific falsification, and probably the majority, arises not because of any nefarious intent, but because of a combination of poor training, poor communication, self-deception combined with a stubborn refusal to admit a mistake. In my view if we want a "punishment that fits the crime", the "crime" is usually a distortion of the research reported in a publication, and the appropriate "punishment" would be a retraction.

There is a crying need for procedures that deal with serious misconduct more expediently than the current system permits. While there are essentially no data on the duration of research misconduct investigations, all of the available examples that we can read about have taken many years to adjudicate. Using retractions as a window on the overall timeline for such a process, a recent study reported the mean time to retraction of falsified articles involving senior authors was 79 months [306]. It is imperative institutions take their responsibility for ethics more seriously if they are to become equity stakeholders in companies, maintain patent portfolios and develop for-profit educational materials. These goals are undermined by federal regulations, which compel universities to investigate their own personnel, an obvious conflict of interest. Unless we can reform the current research misconduct system, we are setting our society up for an ethical quagmire, in which universities, the former ivory towers, may legitimize unethical practices.

Chapter 10
Sunshine Laws and the Smokescreen of Confidentiality

> *"The best weapon of a dictatorship is secrecy, but the best weapon of a democracy should be the weapon of openness."*
>
> *Niels Bohr*

The contradiction between the confidentiality practiced by adjudicating institutions and the public nature of academic science causes disruption of every aspect of research misconduct investigations. Prior to adjudication, allegations would best be kept from public view, but this can rarely be achieved in a collaborative research setting. The difficult problem of reigning in rumors or protecting informants and respondents from repercussions to their careers is often ignored by university administrators, even though the purpose of the NSF OIG confidentiality regulation is to protect the individuals involved. Rather, the result of the regulations is that universities impede the release of documents, including threats from lawyers or conclusions that data have been falsified to protect the reputation of the institution, but to the detriment of good-faith whistleblowers or falsely accused respondents. Universities will usually state that they base their interpretation of the confidentiality regulations on the interpretation by NSF OIG. However, that can be a smokescreen since university General Counsel will be the arbiter of what the federal interpretation is. A faculty member or other observer who does not have a legal background will be dismissed as not having sufficient expertise to interpret the regulation. Since I have conducted a court case based on precisely this issue, I have some of my own training, although I make no claim to be a legal scholar.

NSF OIG research misconduct regulations refer both to FOIA and to the Privacy Act (45 CFR 613.1 *et seq.*) in describing what information can be released by the OIG. One can see the FOIA and Privacy Act as two laws that have diametrically opposing effects. The FOIA provides for release of public records by federal agencies at the request of members of the public. The Privacy Act protects the identity of individuals and requires confidentiality of their personal information in all government procedures and documents. In practice, the OIGs will release documents pertaining to a completed research misconduct case, but only in redacted

S. Franzen, *University Responsibility for the Adjudication of Research Misconduct*, https://doi.org/10.1007/978-3-030-68063-3_10

form. This means that all personal identifiers and other information about individuals is blacked out. Universities are supposed follow these regulations, but their selective implementation of confidentiality can undermine the intent of the regulation. Specifically, confidentiality can be used to defend the privacy of university administrations, while the administration refuses to defend the rights of employees involved in research misconduct cases.

10.1 The Contrast Between Judicial and Research Misconduct Systems

The legal landscape of confidentiality regulation involves the conduct of the investigation in ways that appear at odds with general principles of justice in the United States. The sixth amendment to the U.S. Constitution guarantees a defendant the right to know his/her accuser. Constitutional regard of citizen's rights has antecedents in law dating back hundreds of years. The idea that one should know one's accuser, have a right to know what witness testimony is offered and a right to a fair and speedy process are time-honored concepts. Although research misconduct is not a crime, the severity and duration of a research misconduct case can have negative repercussions on a person's livelihood. In contrast to the legal system, in a research misconduct investigation, a respondent does not have the right to know their accuser. In spite of the clause in the regulation stating that the identities of the informant and respondent must be kept confidential, it is virtually impossible to keep the identity of the whistleblower secret from the respondent. As currently implemented, research misconduct proceedings are definitely not speedy procedures. In practice, witness testimony is conducted behind closed doors so that neither the informant nor the respondent may know what testimony was given. Evidence is most often kept from public view and most often it is even kept from the informant, who is often the most knowledgeable witness. Confidentiality is used to prevent anyone from knowing how a committee decision was reached. The potential for abuse is obvious.

10.2 The Futility of Confidentiality as a Protection for Either Informant or Respondent

Regardless of the intent of the confidentiality clauses in the regulation, it makes absolutely no sense to pretend that the informant can remain unknown to the respondent in a research misconduct case [406]. By definition, research misconduct allegations concern scientific publications. The median number of authors on a high impact scientific publication was found to be 3.5 in a recent study [411]. Fully 80% of the studies had less than five authors. The identities of those who contributed to the work and those who would know enough about the work to make an allegation are known to all parties involved in most cases. While criticism of a publication may happen by way of a posting on PubPeer, bloggers are not usually the ones who would file an allegation. The cast of characters who could possibly know enough about the science to make an allegation is almost always limited to a small number of people who have intimate knowledge of the research either by having conducted it or because they are affected by it as either a collaborator or a competitor in the community. By the nature of the allegation the respondent will be able to presume its origin, someone close to the research who has likely already expressed concern. Despite these facts, the OIG and university RIO will act as though they are protecting informants by keeping their names secret, even when it is obvious that the identities are known. This is an example of the misplaced priorities of the regulation. The agencies spend a great deal of effort to enforce a meaningless procedure, while ignoring the fact that universities can abuse confidentiality to protect themselves from bad publicity. One of the main potential sources of bad publicity is mistakes or improper conduct of research misconduct cases. Universities do not have a good track record when it comes to supporting faculty whistleblowers. Student and post-doc informants fare worse. While retaliation by the respondent is a concern, the greater danger is institutional retaliation.

It is almost amusing the lengths that the OIGs will go to in pretending that the informant/whistleblower is anonymous. I have personal experience with this because when allegations with retaliatory animus were made against me in 2015, I had occasion to speak to the agents at the NSF OIG to understand the history of the allegations. One always has the right to speak with the agents about one's own case, but it is still quite difficult to learn very much since the answer to almost any question is that the information is "confidential" or an agent will "neither confirm nor deny". In my case, I was trying to understand how long the NSF OIG had kept an allegation about me before forwarding it to the university, which turned out to be 3 years. I was trying to determine whether an identical allegation had been sent to the university and dismissed, since I had uncovered evidence of an allegation against me that the university had apparently dismissed 3 years prior. I was further attempting to establish whether the two allegations were the same, which indeed turned out to be the case. As the whistleblower in the hexagon case, I had a long history with the agent and the NSF OIG and it was obvious that the allegations against me had been written by one of the respondents in the hexagon case as a retaliation. Yet, the agent went to great lengths to pretend that I could not know for certain who had made the allegation. The agent objected every time I used the name of the informant, whom we both knew, and insisted that the informant was anonymous. The agent was merely following protocols, albeit absurd protocols. By asking leading questions, I was able to force the agent to reveal I was correct in my assessment that the NSF OIG had decided to complete the hexagon case before determining how to proceed with allegations that were potentially submitted with retaliatory animus. But by this time I had 9 years of experience speaking with the agents and I understood what kinds of questions they could answer how they could admit certain information.

Of course, the intent of confidentiality of informants in misconduct regulation is to protect vulnerable informants from retaliation. For example, a junior person in a laboratory who makes an allegation may be subject to retaliation by a mentor. It is noble of the agencies to want to protect vulnerable whistleblowers, but the pretense that a scientist cannot figure out who has made the allegation is an insult to everyone's intelligence. A central problem is that the regulation provides for only one remedy for compromised science and that is to make an allegation of research misconduct. It is a serious allegation that implies intent and has far-reaching consequences beyond mere correction of the scientific literature. But without any other avenue and with a no mediator or ombudsman as an alternative for resolution, there is no way to step outside the laboratory to another level of scientific expertise to examine certain facts and discuss the discrepancies. Irrespective of whether a graduate student or a senior professor discovers the discrepancy, the only path is to report an observed misrepresentation as "research misconduct" to the RIO of the university.

While it sounds well intentioned to protect both informants and respondents from negative consequences of an inquiry by protecting their names from the public, this is an unrealistic expectation. There are always numerous observers in any laboratory who will know about the discovery of misrepresentations that could potentially be ethical transgressions. There is almost nothing more fascinating to break the humdrum of everyday science than the possibility that someone has done the unthinkable and manipulated data. If data have been made up out of thin air it is altogether titillating to consider why someone might have transgressed in such a way. Therefore, in cases that involve senior scientists, the admonition to guard confidentiality is unlikely to have any weight. The idea that confidentiality would stop tenured faculty from speaking openly about the subject is naïve. In the hexagon case, everyone in the academic departments of both universities involved and many in the university administration knew who was involved in the case from the beginning. I do not condone rumors, but they are a fact of life. In the hexagon case I was circumspect for 5 years, but I quickly learned (from emails in my inbox) that the respondents had told everyone in my workplace about their indignation over the injustice of the allegations and my incompetence. As they were publicly ranting, they were privately filing allegations against me as quickly as they could come up with them. Much later I learned that everyone in the administration from top to bottom in the Graduate School, General Counsel's Office, Chancellor's Office and Office of Sponsored Programs knew about the case and knew about the allegations made against me. The members of the University 1 administration did maintain confidentiality in the sense that they did not inform me of the allegations made against me or their deliberations based on those allegations, but they did not enforce confidentiality when I asked them to prevent my reputation from being attacked by University 2. Moreover, they invoked confidentiality to justify failing to notify journal editors of the finding of falsification by their own investigation committee and later by the NSF OIG. Meanwhile, my reputation suffered in every avenue of my professional life for having "blown the whistle".

The one situation where confidentiality could possibly function as intended is in the conduct of a case against a junior person or graduate student. During the short duration of such a case it is likely that the graduate student respondent would be fearful of speaking out. Unfortunately, the respondents in such cases have few options and if the evidence points to the fact that there are mentoring issues, then confidentiality will only help to protect the mentor. Most graduate students and post-docs who are found to have committed research misconduct are asked to leave the university. Confidentiality does nothing to protect them from this fate. But it does protect the university from having to look into whether faculty should have done more to mentor the student or perhaps even encouraged the student in inappropriate ways.

10.3 The Perverse Use of Confidentiality by Institutions for Their Own Protection

Institutions will usually invoke confidentiality to prevent the release of reports that reveal investigative committee conclusions. While the scientific findings are public, the conclusion that the papers contain falsified data or image manipulation may remain secret until there is a finding of research misconduct. Most universities will refuse to release the final report even after finding research misconduct, although that refusal is contrary to federal regulation. The intent of the FOIA law is that redacted versions of official agency documents pertaining to research misconduct investigations should be made public upon request. It is remarkable that the finding of research misconduct by Wayne State University against Dr. Sarkar was not exposed until he sued PubPeer for permitting an anonymous informant to point out image manipulation in his papers [338]. Perhaps Dr. Sarkar threatened Wayne State with legal action if they were to release the report describing their finding of research misconduct or possibly the university felt that it was a blemish on the institutional reputation. While the NSF OIG is often responsive to properly worded FOIA requests, other agencies are so guarded that they will go to extremes to prevent the release of any information about a concluded research misconduct case. In a rather humorous article in *Nature News* entitled "Misconduct oversight at the DOE: Investigation closed" reporter Eugenie Reich describes a research misconduct case in which officials at the DoE apparently exonerated the respondents without ever reading the final investigation report. The case involved an allegation lodged against the Pennycook laboratory at Oak Ridge National Laboratory (ORNL). The Department of Energy (DoE) issued a public exoneration of the individuals upon receipt of a final investigation report from ORNL. Ms. Reich requested the report using the FOIA mechanism. The FOIA request was denied and Ms. Reich and the journal *Nature News* filed suit against DoE under the FOIA [404]. The defense offered by the DoE lawyers as a justification for their refusal to release the report was that the report had not been made official because no one in the DoE had actually read it. Ms. Reich reports in the article that five DoE officials testified under oath that they had agreed to exonerate the respondents without reading the report.

While this approach certainly lessens the burden on administrators in the short run, it is unlikely to encourage greater ethics in scientific reporting.

10.4 Confidentiality Hinders Adjudication of Research Misconduct

Confidentiality hampers many aspects of an investigation. The members of an investigation committee may feel that it is not appropriate to consult outside expert faculty members about the scientific issues for fear that they may violate confidentiality. This impedes the normal kind of scientific discussion that permits one to organize observations and understand a result. If the committee is dealing with a complicated case of falsification it can be like any other scientific puzzle. One needs to understand many aspects of the behavior of matter or cells or rocks (depending on the field of inquiry) that would inform whether the statements made in a journal article correspond appropriately to the experimental data. One needs to be able to look at the science openly just as with any other type of scientific conundrum.

Witness testimony is usually kept confidential in a research misconduct proceeding. If neither informant nor respondent have knowledge of the witness testimony, there is no way to cross-examine a witness to verify testimony. One must assume that the justification is the same as that for keeping the identity of the informant "confidential". Some witnesses may be in a vulnerable situation and perhaps would be reticent. On the other hand, there are situations where the respondent is the one who is vulnerable. What protections are offered then? If a witness is lying, the committee may not have any way to know. The respondent would be in a much better position to expose the lie by cross-examination than a committee member, but the respondent would not be present for the interview. Nor would the respondent usually see the transcript of the testimony. The current system has serious impediments to meaningful investigation.

Evidence is supposed to be sequestered immediately in a research misconduct inquiry to prevent tampering. This requirement is described in ORI regulations and protocols sent with NSF OIG referral letters. This requirement is often interpreted to mean that evidence must be kept confidential, although it is not explicitly stated in the regulations. When institutions take the further step of keeping the sequestered evidence confidential, they remove the facts of the case from the informant and the witnesses. In the hexagon case, as the informant and primary witness in the case I was denied access to the laboratory notebooks throughout the duration of the case at University 1 (2006–2008). During the investigation I was sufficiently intimidated by the claim that I had no right to see the laboratory notebooks that I did not file a formal request. Subsequent to the conclusion of the investigation at University 1 I began to realize that confidentiality was not working to promote a fair process. When I asserted my right to see the laboratory notebooks, University 1 legal counsel denied my request on the grounds that they were part of a "personnel file", which is exempt from the Public Records act under G. S. § 126-24 of North Carolina. The fact that the reason for denying access had changed from confidentiality to a personnel file exemption made me suspicious that General Counsel was coming up with excuses to prevent me from seeing the evidence. Moreover, the argument that a laboratory notebook was part of a personnel file sounded spurious to me. I retained an attorney to convince the university that I had the right to see the graduate student's laboratory notebooks. When I finally saw the laboratory notebooks in February 2010 the evidence of intentional falsification was clear [30]. I could also see that the committee had not consulted the notebooks. But by then it was already 18 months after the University 1 investigation report had been written. If the NSF had not accepted the case, it would have been too late to use the information. Because of the confidential nature of the investigation, I will never know why the committee did not consult the most obvious evidence in the case. From my conversations with retired administrators (years after the conclusion) I know that the committee never visited the evidence room or saw the notebooks. The federal agents in the NSF OIG consulted the notebooks after February 2010 because I wrote to them to explain the evidence that I had found there. The NSF OIG wrote in their final report in 2015 that the laboratory notebooks provided definitive evidence of falsification and of recklessness. If I had not retained a lawyer and forced University 1 to give me access to the laboratory notebooks, the NSF OIG would have simply closed the case without making a finding. Ironically, the fact that the NSF OIG was so slow in their adjudication actually worked in favor of adjudication in this instance. Although it had taken me almost 2 years from the end of the investigation at University 1 to obtain access to the notebooks, the NSF OIG was still working on the case.

10.5 Public Records Lawsuits as a Remedy for Violations of Whistleblower Rights

My experience with public records law can hopefully be of value for good faith whistleblowers who confront retaliation. Public records can serve as evidence of improper actions by a university administration. Requesting those public records that demonstrate improper administrative actions can serve as defense against harassment and retaliation. In cases where whistleblowers have received poor treatment by their institution, they may end up contemplating the legal action. People often consider defamation lawsuits, but these are unlikely to succeed, and they force the plaintiff into the role of aggressor. This can defeat the purpose of trying to find a path forward to resolution of an institutional grievance even if the case is technically justified. If a whistleblower is aware that administrative actions that are to his or her detriment, public records lawsuits are an appropriate complaint since the records being withheld by a university often will reveal an injustice. The university is on weak ground since the administration has no legal right to withhold any part of a personnel record of an employee from the employee him or herself. Documents that pertain to treatment of the employee, who has blown the whistle are part of the employee's personnel file. Documentation revealing incidents that may have negatively affected the whistleblower, such as threatening letters from lawyers that attempt to discredit the whistleblower should also be part of the personnel file or, at least, they should be public records. Although the threat of bad publicity is one driving force towards a mediated settlement, perhaps the strongest motivation for a university to negotiate is a fear that release of public records could result in greater legal problems.

Most states have whistleblower laws, but these apply only to state employees. Professors and post-docs at public universities are state employees so these laws may offer some protection. Unfortunately, students are usually not considered employees. The system disfavors student rights at every turn. The whistleblower statutes were written with financial misconduct rather than academic misconduct in mind. This fact may make it more difficult for a research scientist to use the state general statutes as part of a complaint. Nonetheless, most whistleblower statutes define both harassment and retaliation as abuse. While retaliation usually refers to termination, administrative actions such as research misconduct inquiries against a whistleblower may well constitute harassment. Hence, a research scientist who has not been terminated but has suffered because of poor treatment may still be able to demonstrate harm, which is required in order for a court case to proceed. While public records law does not actually require a demonstration of harm, there is a minimum threshold of damages for a filing in Superior Court. Thus, under the common scenario where a whistleblower suffered some form of retaliation in the workplace, it is a sensible legal strategy to combine two separate sets of claims, a public records complaint and a claim of retaliation or harassment against a whistle-blower under a state personnel act [412].

There is no statute of limitations for a public records lawsuit. This is important since the statute of limitations is a crucial obstacle to many types of lawsuits. For example, there is a statute of limitations of 12 months for any whistleblower complaint. The State Attorney General's office, who represented University 1, claimed that the statute of limitations is 12 months from when the alleged retaliation was initiated, not from the time it was terminated. This detail shows how subtle interpretations of the wording of the statute (and precedent) may determine whether the case is dismissed or not. The two claims were separate so even if the retaliation claim had been dismissed, the public records part of the case would have proceeded [382]. Although the damages in a public records lawsuit are modest by legal standards, there will normally be an incentive for a university to negotiate. Aside from the possible bad publicity, the potential for release documents showing poor treatment of a whistleblower may open up the university to further legal liability since the statute of limitations on any new claims resulting from the documents begin when the documents are released.

10.6 The Confidentiality Smokescreen Protects the University Administration

As a result of the implementation of FOIA in the agencies and the other considerations at universities, there are few statistics and little information available for public policy discussions. Indeed, any attempt to obtain even the smallest piece of information will involve a protracted legal battle. Newspapers have historically been the plaintiffs in most public records lawsuits. If science were as interesting to the readership of newspapers as embezzlement or sex scandals, there might be some hope that newspapers would champion public records lawsuits against universities to obtain some of the information that would help to understand the issues in contested cases. However, legal threats to prevent the retraction of falsified research articles are unlikely to inflame public opinion. Without hard evidence in the form of withheld documents, there is little a reporter could write about. How will we ever reform the process if we do not know which factors are leading to contested or controversial decisions?

Honest scientists should welcome an open process. If a scientist is wrongly accused, that will emerge during the investigation and everyone will know. If a scientist makes a frivolous allegation that is also something for which he/she should take responsibility. An open process is more likely to be fair to all concerned and it is more likely to be a speedy process. Many scientists lament that fact that students are often scapegoats for misrepresentations in published work where there was a collective decision made to publish. But confidentiality is not protecting the more vulnerable junior scientists either. In fact, confidentiality can be used as a shield to protect the professors and the institution in such cases. Universities can revoke the rights of foreign-born graduate students to stay in the U.S. For international students there may be no time for an appeal or independent examination of the facts. Even for domestic students, an examination of the facts kept confidential by the university is only possible by filing suit in a court of law. Victimized students do have such recourse, but it would be exceedingly challenging to exercise that right. Furthermore, even when right is on the side of public scrutiny of the records, the legal battle required to establish that right is beyond the means of most students.

There has been an enormous investment in our educational system and yet there is also a high price for access, which affects every parent who plans to send children to college. Standards for ethical behavior should be of concern to the public at large since the maintenance of these standards is essential to the training of students and ultimately the soundness of the entire university system. It is possible for a research group to encourage a student to cherry pick data and selectively examine the facts to substantiate a hypothesis; then the student may be blamed for absorbing the expectations of the group. While many scientists may bristle at the suggestion that this happens, the frequency with which students are found to have selected data or selectively altered images strongly supports the hypothesis that this behavior is learned in some research groups. The surging retraction rate and reproducibility crisis both point to systemic ethical issues. The secrecy of university research misconduct processes threatens the fabric of our educational system since we have no independent way to evaluate either the ethical failures or university investigations that are supposed to adjudicate them. Students and the learning process are both hurt when universities hide the fact that blame for ethical transgressions is frequently shifted to those with the least power and the adjudication of such matters is subservient to financial interests.

Chapter 11
The Legal Repercussions of Institutional Conflict of Interest

> *"Now, imagine you're a researcher who wants to game this system. Here's what you do. Run many small and statistically weak studies. Tweak your methods on the fly to ensure positive results. If you get negative results, sweep them under the rug. Never try to check old results; only pursue new and exciting ones. These are not just flights of fancy. We know that such practices abound. They're great for getting publications, but they also pollute the scientific record with results that aren't actually true."*
>
> *Ed Yong, The Atlantic, Sept. 21, 2016*

As public sources of funding have become more competitive, there has been greater pressure to accept funding from private corporations [413]. Private university initiatives may be internal, based on university patents, or from external sponsors. For example, academic departments at public universities are all but forced to create educational materials to sell to their own students to subsidize their operating budgets. New software, web-based tools and instruction are created and copyrighted as part of this effort. Patenting, and participating in start-up companies may come into conflict with the training of graduate students. Nonetheless, the incentive to generate potentially lucrative intellectual property drives many professors to patent discoveries or copyright educational innovations. While the genie is out of the bottle in terms of university patenting and ownership positions, one would hope that there is a role for ethics in the burgeoning technology transfer industry. However, in the tech transfer world, there are few ethics workshops or courses for would-be academic entrepreneurs. Instead, the backstop is the research misconduct system. Although a great deal of attention is paid to faculty conflicts of interest [414, 415], the role conferred to universities by the Bayh-Dole act generates institutional conflicts of interest that have only recently been considered in the debate on research misconduct policy [416–418]. However, if an ethical issue arises in the process of patenting technology and marketing it to investors, it seems obvious that the university cannot properly adjudicate a potential ethical breach in its own investment. Yet, current research misconduct regulation does not provide any other option.

S. Franzen, *University Responsibility for the Adjudication of Research Misconduct*, https://doi.org/10.1007/978-3-030-68063-3_11

The trend towards permanent budget cuts by many state legislatures has forced universities to redefine their funding sources and even their role in society [227]. Universities may become the owners of technology patented by faculty and the direct beneficiaries of the sale and licensing of patents. Despite the importance of maintaining a reliable record of research for patenting and discovery of new technology, there are documented cases where university administrators have failed to act in the interest of the public good by neglecting to notify researchers, editors, or other affected parties when falsification or fabrication of data has occurred [419] Institutions may sometimes refrain from an assiduous investigation of possible misconduct, however meritorious the allegations, for a variety of reasons including the following: The possibility of the university's own financial loss [420]; fear of bad publicity [421]; or fear of a lawsuit [338, 422]. Since patent portfolios and spin-off companies are at the core of a new funding model there is a great reluctance to do anything that might jeopardize the university's strategic position.

11.1 The Pragmatic Approach to Ethics in Modern Universities

Even if administrators have the best of intentions, under the present research misconduct regulation, universities have an incentive to do the minimum necessary to satisfy the federal agencies. Research misconduct investigations are both expensive and divisive. They pit faculty against other faculty and/or students. They can result in bad publicity or even lawsuits. Moreover, experienced administrators will know how difficult it can be to conduct a contested case to completion given the burden of proof of "intent" or "recklessness". To the extent that alternative solutions have involved bringing in an ombud or a mediator to try to resolve certain issues it could be a healthy development since one of the weaknesses of the regulation is that it offers the sole option of an investigation into allegations of research misconduct. On the other hand, the combination of laissez-faire oversight and the economic incentives inspired by Bayh-Dole, underscore a drive towards privatization that has put enormous pressure on universities to be pragmatic about ethics. The lofty goal of becoming the economic engine for development can override more mundane concerns regarding publishing standards and the accurate record keeping of science. The need to generate revenue to support the teaching mission justifies expediency in the conduct of expensive investigations. Precisely at a time when ethical issues are expanding because of the new pressures of commercialization and reduced public funding, the universities are under financial pressure to reduce the expense of lengthy research misconduct deliberations. Neither the public nor the scientific community has any way to gauge the ethics of the university since all these deliberations happen behind closed doors.

Even when research misconduct cases are prosecuted to completion by federal agencies after many years, a finding of research misconduct does not necessarily lead to correction or retraction of affected publications. The mandate of the federal agencies is to protect the financial interests of the taxpayer by exposing falsification and fabrication, but the goal of correcting the scientific literature is delegated to the universities. Although the OIGs and ORI have an abstract interest in seeing the research literature purged of falsified data, their regulations do not permit them to play a role in that process. The OIGs/ORI do not admonish universities to take action either, unless there is a gross procedural violation. A recent article by members of the review committee responsible for oversight of the Duke University handling of the Potti case criticized the ORI for failing to make a minimal commentary of the poor institutional response by the Duke Medical School administration [423]. Even though telling the truth in research has historically been the *raison d'etre* of the university, in the absence of regulatory insistence, universities rarely admit that their faculty is responsible for infelicitous contributions to the published research record that need correction.

Given that fundraising has become essential for universities [424] to pay for quality facilities and compete in the marketplace for quality faculty and students [425], even a senior scientist can fear reprisal from the university administration for bringing forth allegations that risk damaging those activities. Being a whistleblower has always been an "unpleasant avocation" [346], but it is increasingly the case that reporting ethical failings will be directly adversarial since it can affect the bottom line of the institution. The fact that the whistleblower may be correct and possibly save the university from gross negligence by preventing support for falsified research is not likely to bring accolades from the university administration. For one thing, the administration is likely to hold the whistleblower in at least as much suspicion as the respondents until a case is adjudicated. By that time, the whistleblower will likely have left the institution because of negative fallout of colleagues and administrators who only know what is in the rumor mill. Rumors are not usually a whistleblower's friend. When money is involved tension flare. Evaluating the truth of the allegations can be exceedingly difficult when the complex nature research results from multidisciplinary team is muddied by the perverse incentives of commercialization.

11.2 The Covert Power of Legal Threats in a Research University

Legal issues surrounding corrections to the research literature are an increasing part of the consideration of any debate about scientific integrity. University legal counsel will be consulted as soon as the respondents or informant retains a lawyer, and sometimes beforehand. If the outside lawyers are actively trying to question or delay the proceedings, then university legal counsel must write briefs to justify university procedures. If university general counsel spends time on the case it is likely that the upper administration will know, i.e. the Provost and Chancellor. Any significant legal threat is likely to be discussed at the highest levels to coordinate the response given the high stakes for the institution. Thus, when facing legal challenges against the university, the confidentiality of the whistleblower is not protected. In fact, the whistleblower is particularly vulnerable to decisions from the highest ranked decision makers in the university. These individuals are often removed from the facts of the case and unfamiliar enough with the science that they must rely on outside experts to make any case for image manipulation or fudging the data. Simultaneously, confidentiality leads to compartmentalization of the case. This means that the evidence and facts of wrongdoing are not known, but the legal threats and counterclaims by respondents receive the full attention of the decision makers. The fog of confidentiality can hurt all aspects of decision making concerning the case and its aftermath since it is difficult to know if one has all the facts. High-level administrators may not know whether a case has merit, but they will know when there is a potential for loss of private investment, a lawsuit or damage to the prestige of the institution. In this context, it is easy to misunderstand facts, succumb to a narrative developed by lawyers or for doubts about the whistleblower's motives to cloud the judgement of university administrators.

The hexagon case has provided some clues as to what kinds of legal strategies might be used by respondents who want to shut down an investigation. Respondents can object to procedures based on defamation law, but intellectual property law is much more powerful where it is applicable. Patented technology can serve as a reason to restrict usage or access to certain products or methods, but also to scientific data. If the claim is made that data under investigation are part of an intellectual property portfolio, it can be difficult for a university to prove otherwise. If the data are proprietary, the respondents may claim no one has a right to samples or materials used to investigate whether the data are falsified. Intellectual property law can also be invoked to prevent publication of scientific data by claiming that the data are proprietary. A company associated with the patent must give permission prior to publication. The point of such legal threats is not to insist on patent rights, but to conflate patent rights with ownership of data and thereby to intimidate editors or administrators to prevent critical publications or investigations from proceeding. While the threats would most likely not stand up in court, no one wants to call a lawyer's bluff. University general counsel has a mandate to keep the university out of court so the threat of taking the university to court often has teeth just because any court case, no matter how frivolous, is to be avoided at all costs.

While defamation lawsuits are common threats issued by respondents, this type of lawsuit is very unlikely to succeed in a research misconduct case. Judges are not typically sympathetic to the plaintiff in defamation cases and the burden of proof is on the plaintiff. In a defamation case the speech must be false, it must directly impugn a person's character and one must prove that the false speech has directly resulted in financial damages. Proving harm from research misconduct in a courtroom is difficult. In reality, harm to one's reputation can be devastating in science. But proving it is another matter entirely. The reviews of grant proposals and manuscripts that one receives are anonymous. The fact that one's rejection rate increases dramatically could be mere coincidence and it would be difficult to prove that it was because of the harm caused by defamation. In legal cases, one must distill the harm to a dollar figure, which is even more difficult when it comes to a matter of reputation.

Intellectual property law, on the other hand, offers a way for the respondents to threaten a lawsuit while also portraying the informant as an aggressor. Further, it may be claimed that no one has the right to publish any conclusions related to the data since they are proprietary. Indeed, if a Material Transfer Agreement (MTA) or other specific agreement exists then there is legal restriction on the use of the data. In the hexagon case, there was no MTA or agreement of any kind. This did not stop lawyers for a company in State 2 that employed one of the respondents from writing threatening letters to University 1 General Counsel claiming that our research group had infringed on its patents by publishing a scientific paper in *Journal of the American Chemical Society*. University legal counsel may realize that the alleged patent infringement is not a valid argument for challenging a scientific journal article, but they will nonetheless seek to placate the attacking attorney to avoid any risk of escalation. One might think that such arguments would be invalid in a university due to public nature of research and the lack of profit in the pursuit of knowledge. However, it is important to realize that the Bayh-Dole act has changed the definition of patent infringement in universities.

11.3 The Vulnerability of Post-Bayh-Dole Universities to Intellectual Property Lawsuits

The application of patent law to a non-profit institution or university first became part of patent law with Federal Appeals Court decision in Madey vs. Duke University in 2002 [426]. Patent law pertains to the use of technology for financial gain. Under this definition, one infringes on the patent only if one produces or sells a patented product and then specifically when one makes a profit from that activity. For many years, universities maintained that they had non-profit status and that they were exempt from patent laws. However, this idea changed when a dispute between an inventor and a university led to a court case concerning patent rights. Prof. Madey was the inventor of the free electron (FE) laser, and Duke removed him as the director of an optical science center in a disagreement over use of the FE laser. He left the university, and then sued Duke for patent infringement over their use of the FE laser for biomedical research. Duke lost the case in court because the medical school could not get the judge to accept the argument that it was a non-profit organization. Duke receives hundreds of millions of dollars a year in revenues from patients who receive treatment at its hospital and various spin-off companies that have developed new products in the biomedical arena. In years past, the patent exemption would probably have saved Duke University, but after the Bayh-Dole act, universities that are engaged in profitable activities can no longer make a convincing case that they are acting as non-profit organizations.

After the decision in favor of Madey, the use of patented processes in the university is only permitted if the experimental use provision holds, namely if the use is "solely for amusement, to satisfy idle curiosity, or for strictly philosophical inquiry" [427]. The definition of philosophical inquiry is crucial, but its interpretation is not particularly clear in the modern context. The original phrase introduced in an 1813 decision by Justice Joseph Story probably meant something quite different then than it does today. In practice, universities have claimed that they are engaged in philosophical inquiry and provided they are not engaged in related activities that generate revenue for the university this a defensible position. The application of Madey vs. Duke raises questions for the future since that decision alters the rights to use technology, at least in private universities. Madey vs. Duke strengthens the hand of anyone who wishes to use intellectual property law to intimidate others. The net effect of the legal threats using these provisions is to restrict open access to evidence sequestered in a research misconduct investigation, which is the effect desired by those who have something to hide. Legal intimidation based on intellectual property law has a chilling effect of the pursuit of a research misconduct case or the willingness to notify journal editors of a finding of falsification [382].

11.4 The Situation of Private Universities

The highest ranked private universities may be able afford greater attention to research misconduct issues, particularly in the hard sciences and engineering. The top tier university also have a status to uphold, which drives them to treat integrity issues with greater care than state institutions. They have a brand to protect. Moreover, top tier private universities are buffered by enormous endowments, which tends to mitigate the financial loss of a finding of research misconduct. On the other hand, the conflict of interest issues in the top medical schools have become a significant ethical issue in top tier universities. All of these universities have had to confront the fact that their faculty can make substantial consulting fees and there are always pharmaceutical and medical device companies ready to enter into agreements that risk comprising the objectivity of those faculty [245]. The private institutions are in a competitive market for students, which also drives much of their decision-making. Thus, they are more likely to be concerned about the perceptions of those students and their parents than the public institutions are. A private university must take greater care to ensure integrity in order to be able to present their faculty as worthy of the high price of tuition. In an article entitled "Harvard Medical School in Ethics Quandary", Duff Wilson reports on how students and faculty at Harvard Medical School have exposed the degree to which the faculty there have received industry support, with numerous examples showing the influence on the statements and judgements of the faculty. Harvard received a F letter grade by the American Medical Student Association [428]. By contrast, state universities are subject to enormous budgetary pressures, which has led to acceptance of industry funding from agribusiness, oil companies as well as pharmaceutical companies and many others. Although different in the locus of potential problems, no educational institution is immune from ethical issues associated with the increasing commercialization. While the top tier private institutions may have the resources and perceived stature to treat issues of integrity with greater care than state institutions, both public and private universities feel the pressure of intellectual property concerns. These concerns can steer the institution and have profound effects on the outcomes of any contentious ethics case.

11.5 University Conflicts of Interest

There are three types of ethical conflicts that can be affected by university financial relationships. First, spin-off companies represent a relatively new phase of development in the history of the university, but one of increasing importance. Technology transfer offices need to assume a greater responsibility in monitoring the reporting of research. Second, clinical trials or other product safety trials in the development of a new drug or diagnostic can lead to concerns over objective reporting [123, 429]. There have been serious ethical errors and intentional disregard for standards in clinical trials at several university medical schools. Third, research into the toxicology of industrial compounds or ecological consequences of energy technology, agriculture etc. can result in institutional conflicts of interest, particularly when they are supported by industrial partners. When companies are major donors to the university it can be difficult or impossible to publish scientific papers that document health or environmental effects arising from their activities. Examples include research into fracking, tobacco research, climate change research, as well as research on pesticides and food additives [4].

Increasingly, university administrators court corporate funding of research to compensate for decreased government support. The ensuing relationships may abuse the mission of the university. Research judgments will naturally align with the priorities of the sponsors, even in the absence of an explicit agreement to do so. For example, a link between a fracking industry group in Texas and a research scientist at the University of Texas in Austin called the conclusions of a recent study into question [430]. The study authors concluded that there is no possibility for groundwater contamination by fracking. Subsequently, it was revealed that the research had been supported by more than a million dollars of funding from the industry group. Many academics believe that corporate funding inherently will undermine the independence of academic research [413]. The amounts of corporate funding at most universities remain confidential, and as we have seen, university administrative policies are well crafted to preserve confidentiality.

A high-level view of the peril of industry-university partnerships can be found in "Corporate funding and conflicts of interest - A primer for psychologists", which presents the unanimous findings of the American Psychological Association (APA) task force on external funding published in *American Psychologist* in 2007 [431]. There was great concern regarding the close contacts between industry groups and academic administrators, the lack of oversight or disclosure. Often faculty are suspicious of such funding arrangements, but increasingly they are being encouraged to participate by the administration because of the tight funding climate. When funding options are sufficiently narrowed by the reduced support from government sources, some professors simply have no choice.

The influence of biotechnology companies on research in universities has been especially controversial. When research produces findings that are not convenient for the corporate sponsor there are always other research groups, anxious for funding, that will find agreement with industry expectations. In her book, "University, Inc.", Jennifer Washburn writes about the "experiment" at U.C. Berkeley in which Novartis and the Department of Plant and Microbial Biology established an institutional agreement that gave Novartis access to researchers and their results in a less restricted way than typical corporate research sponsorship [4]. Supporters pointed to the advantages to the department because of access to the genetic database owned by Novartis in addition to the large infusion of research funding. The blanket agreement for $25 million, which gave the company access to all of the research in the department, was in force in 2001 when Dr. Ignacio Chapela published a study in *Nature* showing that genes from genetically modified corn had been found in native strains of *Zea mays* in Oaxaca, Mexico. Mexico does not permit the use of genetically modified crops and regards the native strains as essential to its agriculture [432]. Chapela's research group also reported that the modified strains had been able to reproduce and spread [433], which was ultimately found to be incorrect and led to retraction of the paper. Neither Nature nor University of California at Berkeley alleged unethical behavior. Since Dr. Chapela had been outspoken about the risk of genetically modified strains for agriculture in Mexico a scandal erupted in 2003 when Dr. Chapela was denied tenure after receiving near unanimous support within his department and college [434]. However, more importantly Dr. Chapela had been a critic of a $25 million fund provided by Syngenta to various biology departments at UC Berkeley. Such corporate-university arrangements often lead to conflicts as corporate sponsors favor research that aligns with their business interests rather than research on the ecosystem or alternatives to their particular genetic modifications to inhibit insects [432]. Dr. Chapela filed suit alleging that there was a political motivation behind his termination. He ultimately received tenure.

A second controversy at U. C. Berkeley involved the weed killer atrazine [435]. This substance is banned in the European Union, but not in the United States. Based on research funded by Syngenta, Dr. Hayes reported that atrazine caused a change in the sex of frogs, a finding that would suggest a strong hormonal effect. Atrazine appeared to be a so-called endocrine disrupter; a finding that may be linked to declines in frog populations in the U.S. Dr. Hayes alleged that Syngenta started a campaign to discredit him. Syngenta found another partner, Dr. Simpkins. Dr. Simpkins conducted research that showed no effect on frog development. By now a great deal of research has been published that counter Dr. Hayes' findings. Part of the problem in determining whom to believe is that the company has vast resources and once they pull the plug on an academic project that has given disturbing news, it is often difficult for the professor to fund follow-on research.

A third example from the agribusiness sector concerns the effects of genetically modified corn on insect populations studied by Dr. Angelika Hilbeck and funded by Ciba-Geigy (now Syngenta) while working at the Swiss research institute, Agroscope [436]. The corn contained genetic modifications to target insect larvae that would feed on it. Her research found that the corn also killed lacewing, an insect predator considered to be a valuable part of the ecosystem. Ciba-Geigy pressured Hilbeck not to publish the result. When she did so, her contract was not renewed. Subsequently, the company funded another scientist who systematically repeated every study she conducted and came to diametrically opposing conclusions. An outsider cannot prove that this was done by distorting and falsifying conclusions, but the facts strongly suggest that the company disregarded scientific ethics for the sake of product development.

Food and Water Watch published a report on the effect of Agribusiness funding on America's land grant university research programs [436]. The report notes that USDA funding for research fell 12% in real dollars between 2000 and 2010. Agribusiness sponsors have filled at least part of the void. Funds have gone to pay for buildings, endowed professorships and, of course, academic research. The report gives numerous examples of research that has supported products, such as studies showing that drinking soda has no effect on obesity or a study on the benefits of eating chocolate. The results from these studies may be used directly in corporate advertising. The ethical problem is obvious since the companies have a vested interest in the outcome. There is almost no way to separate the for-profit mission of the companies from the funding they offer. When university administrations pressure faculty to take the funding and there is a subtle but pervasive pressure to publish results that favor their product. Such behavior forces researchers into ethical dilemmas. While there is an effort made to write contracts that will protect research, even the best contract cannot change the fact that the university administration will feel strong pressure to minimize discord and bad publicity to maintain a good relationship with a sponsor. The corporate-university nexus has compounded conflict of interest, the conflict of the researcher, of the university, and of the company.

11.6 Universities Adjudicating Their Own Ethics: The Fox Guarding the Henhouse

Some research misconduct will never be considered as such when universities prioritize their funding over research integrity. Often those charged with boosting finances and those responsible academic integrity overlap in the same individuals: deans, provosts, presidents, and boards of regents. If an institution is complicit in unethical behavior, how could it possibly permit an unbiased inquiry into that behavior? One startling case involves misrepresentations to the NSF by a professor, Larry R. Dalton, at another major American public research university, the University of Washington (UW) in Seattle. This case is well documented in the following publication based in part on a public records lawsuit (https://www.jospi.org/article/7765-massive-faculty-donations-and-institutional-conflicts-of-interest): [437], As a Harvard PhD student in 1967, Dalton forged a scientific paper. He copied words, numerical data, and a graph from other scientists and submitted them for publication. This publication was quickly retracted for evident malfeasance. Public records indicate that some at Harvard recommended Dalton to Vanderbilt University as an assistant professor despite conduct that should have disqualified anyone from an industry based on trust. The Vanderbilt faculty were not informed that they were hiring a scientist with a record of fraud. Thirty year later, Dalton's former Harvard mentor, vice provost at the UW, recruited his student to join him in Seattle. In other words, faculty members wittingly recruited a person who had committed fraud to a public university that is one of the perennial leaders in winning NSF funding. Why?

Dalton is wealthy and promised to gift to the UW Department of Chemistry $40 million. Over the years, he contributed between $15 and 20 million, the receipts for which can be found in the aforementioned paper's supporting information [437]. The university also received millions of dollars in intellectual property agreements from a now defunct company, Lumera, built from extraordinary scientific claims.

In 2000, Dalton claimed to have invented a revolutionary new optical circuit element that was certain to "dominate the 21st Century" as told to the *Seattle Post Intelligencer* at the height of the dot com bubble. He stated that his invention would accelerate the internet by more effectively and more rapidly converting electrical signals into optical signals. The innovation claimed was a remarkably strong organization of dye molecules in thin films. Dalton claimed that his research group had gained unprecedented control over the orientations of these special functional dye molecules. He assembled a large team of scientists to compete for major funding from the NSF and elsewhere. There is vast literature on the organization of molecules in thin films and there are simple, inexpensive methods for evaluating the molecules in such films. Despite the fact these simple assays were not performed, Dalton won an $18 million award from the NSF on the strength of his claims. Such a large award is typically distributed to many research groups at the awardee institution. It is also a plum for the administration. One new member of this team, Professor Bart Kahr, the author of the foreword to this book, on discovering that the most basic control experiments had never been executed, attempted to fill in the gaps. However, he and his students could find no evidence for claimed alignment of molecules. As a member of the center, Kahr wrote a report intended for the NSF, as he was required to do. His report explained that claims of molecular orientation, the basis for extraordinary device performance, could not be supported and that federal funds were being ill-directed. However, these observations were not included in any of the reports to the funding agency describing the results of the funded research. Kahr repeatedly expressed his concerns but was advised in writing by university officials – foremost Dalton's PhD mentor -- to stop asking so many questions because he was making other members of the team uncomfortable [437]. At the end of the five-year (2002–2007) funding period, the researchers led by Dalton, applied for a second five-year award for an additional $18 million. The faculty involved in the writing the renewal did not include Kahr's concerns in their reports to the NSF.

Independent charges of scientific misconduct were directed at Dalton's management of the center. A university dean, Ani Mari Cauce, the eventual provost and president of the UW, insisted that Kahr initiate a formal misconduct charge against Dalton. Kahr refused citing institutional conflicts of interest in the form of millions of dollars in donations. Cauce, on her own initiative, and launched two investigations. Her counsel in the matter was the no less than the Attorney General's Office of the State of Washington. Public records show that Cauce was actively soliciting donations from Dalton at precisely the same time she was adjudicating her misconduct investigations. Emails reveal her joking about it [437].

Because the university was delinquent, undoubtedly because of its financial involvement, Kahr asked the NSF to intercede. The NSF OIG dismissed all allegations of a researcher who began his career as a forger, but it refused to release any records to justify its judgment citing exemptions to the Freedom of Information Act including "pre-decisional agency deliberations" and "potential disclosure of personal information in law enforcement files". If the university decides that it is ethical to suppress a report that exposes flaws in a research project, then the NSF will often accept that judgment. This example shows, more clearly than any other, how federal regulations permit the fox to guard the henhouse. According to CFR 689.4, the NSF relies on universities to maintain the integrity of research. Kahr insisted that in this case the NSF OIG must not rely on communications with the university because of its titanic financial conflicts of interest stemming to the large donations it had solicited and received. Consistent with their regulations, the NSF stated: "Improper investigation of allegations by a grantee...is not an issue... that our office would investigate." [437] Federal regulations state that the NSF OIG relies on the universities to maintain the integrity of research. If a university does not do so, there is no remedy in current regulations. The NSF OIG may be consistent with its regulations, but those regulations promote unethical behavior.

Public records requests both the through the State Public Records Act (4 requests), and at the NSF through the Freedom of Information Act (3 requests), produced not one communication between the UW and the NSF. At present, there have been numerous calls by ethicists to make scientific misconduct investigations public [417, 438, 439]. Presently, they are shrouded in secrecy.

Today, the NSF has taken the position, in a letter signed by its General Counsel, that no fraud has ever been ascribed to Dalton, an extraordinary communication in that forgery, plagiarism, data theft, and their adjudication by an admission with which Dalton began his career, are all in the open literature. The General Counsel has further taken the position that because Dalton is free of fraud, by their declaration, he creates no conflicts of interest for his former students in high ranking positions at the NSF. Moreover, the General Counsel has advised Kahr that it is "inappropriate" to discuss their actions, which at the very least indicates that giving comfort to a science fraud is for them discomforting.

11.7 The University Lobby

Universities lobby for their interests just as other businesses do. Lobbying efforts by universities now are 20% larger than those of the coal and gas industry. Universities are gaining freedom from oversight, which permits them to attract students without disclosing their graduation rates [440]. Universities also lobby by establishing relationships that have a direct role in increasing their chances to receive major funding [441]. Some of the richest private universities have the resources to compete with corporations in the patent arena. Carnegie Mellon has used its strong patent position to successfully sue private companies for infringement. The university has been dubbed a "patent troll" for its aggressive stance [182]. This type of aggressive business-oriented activity may benefit the bottom line of the university, but it is hardly what the Bayh-Dole act envisioned when it permitted universities to own IP. By contrast, many state universities are struggling to keep up with the cost of patenting and will likely suffer under the new America Invests Act, which gives rights to the first inventor to patent. Rushing to patent has many of the ethical risks as rushing to publish, particularly in a resource poor university. All the pressures of the business world and intrigues of the political world are finding their way into the offices of university administrators.

The integrity of science depends on calm reflection and careful experimentation to verify results. Of course, full disclosure is crucial in science. Other scientists must be able to review the data and conduct their own experiments that either corroborate or refute the findings. The numerous well documented examples of ethical conflicts between scientists and universities show that universities can define jurisdiction to suit their interests [111]. University 1 used the whitewashed reports from University 2 to try to shut down any further consideration of the hexagon case, despite a finding that data were falsified [42]. There are many cases where postdocs and students have received the blame for image manipulation or sloppy reporting without adequate consideration of the role of senior faculty. Finally, public health concerns have taken a back seat while universities have catered to corporate interests. The license of universities to adjudicate the research misconduct of their own faculty is fraught with many of the problems of associated with the private investigation of business fraud [442]. There are many reasons that in-house investigations fail, but in business at least ethics require professional private services who are supposed to bring appropriate training to the cases they investigate. In universities there is no independence. The RIO, investigation committee, respondent and informant are most often employees of the same institution. While the RIO should have some professional training (although this is not required by the regulation) the committee members are usually untrained in any of the administrative or ethical issues surrounding the conduct of a research misconduct investigation. According to the National Academies "a potentially significant proportion of RIOs were not adequately prepared to fulfill their responsibilities." [443] Scientists may have lobbied for our current misconduct system to avoid the prospect of "big brother" government regulating academic research laboratories, but the compromise that they reached has already failed. We desperately need to find a way to put scientists in charge of adjudicating their ethics in a setting that is both fair and efficient.

Chapter 12
Bursting the Science Bubble

> *"Improving academic morality should be the first step to prevent dishonest academic conduct. One of the concrete reflections of academic moral sentiment is a scholars' inner punishment attitude of shame and condemnation after they implement dishonest academic conduct. Researchers found that inner punishment towards dishonest conduct would predict the incidence of dishonest academic conduct"*
>
> *Sun and Tian, "Dishonest Academic Conduct: From the Perspective of the Utility Function" Accountability in Research, 2016, 23, 139*

Bayh-Dole is part of history and has become a global trend. The transformation has been underway for 40 years. If the scientific community wants to preserve honest and open investigation of science, our mission now must be to reinvigorate universities with a new sense of responsibility to society to be good stewards of technology. We need institutions that can promote ethics on a scientific basis, rather than as an exercise in law enforcement.

12.1 The Unfunded Mandate for Research Ethics

One hypothesis of this book is that poor construction of research misconduct regulation is principally to blame for the crisis in adjudication of research ethics. I assume that the scientists and university administrators have good intentions and that most are coping to the best of their ability. As I got to know some of the people who managed the hexagon case I came to understand the pressure that they feel. Of course, I am referring to meetings that occurred well after the case had been decided and even after the university personnel had either retired or moved on to different employment. There are many people in the field of research ethics who are doing their best, but the regulations are vague, hampering effective administration. The idea that there are good and bad apples, also misses the point. The awareness of fraud is

S. Franzen, *University Responsibility for the Adjudication of Research Misconduct*, https://doi.org/10.1007/978-3-030-68063-3_12

within each of us. More than education, we need a system of checks and balances that keeps honest people honest. Jennifer Crocker wrote: [444]

> "Every minor transgression — dropping an inconvenient data point, or failing to give credit where it is due — creates a threat to self-image. The perpetrators are forced to ask themselves: am I really that sort of person? Then, to avoid the discomfort of this threat, they rationalize and justify their way out, until their behaviour feels comfortable and right. This makes the next transgression seem not only easier, but even morally correct."

The federal regulations do not provide appropriate incentives either to individuals or to universities to participate in adjudicating ethics. Simply put, ethics is an unfunded mandate for universities. I do not see a path forward by writing more rules to attempt to force the universities into compliance so as to achieve the goal of maintaining the research record. One more ethics course to remind students of something that they should already know, is not going to make a difference. Instead, I believe that the creation of a scientific institution for evaluating and administering ethics would have tremendous benefits. I will describe how such an institution would function like an insurance company supported by all academic research institutions, which would spread risk from ethical lapses. It would also function as an ethics clearinghouse, developing uniform policies, maintaining appropriate ethics educational materials, and adjudicating cases for first offenders or minor infractions. Its only mode of punishment would be a forced retraction of an article that violates ethics. The woefully understaffed OIGs would remain in place for the more serious cases that have clear financial aspects because of the magnitude of the abuse. Perhaps the serious cases would receive appropriate attention once the less serious cases are removed from the current OIG case load. Highly contested cases such as the hexagon case could be adjudicated in a reasonable time, instead of the 9-year duration, to which I was subjected as a whistleblower and subject of multiple forms of retaliation.

12.2 The Need for Checks and Balances in Ethics Regulation

The current regulation, which took shape and was implemented from 1985 to 2005 places primary responsibility with universities or "awardee institutions". The OIGs and ORI have the ultimate adjudication authority but are overwhelmed by the current case load and cannot possibly provide adequate oversight. Without checks and balances that make institutions accountable there is a danger that some of them will define ethics in a way that suits their interests, rather than those of science and public welfare [445]. If not universities, who can possibly adjudicate something as complex as research ethics? I will argue that a non-governmental regulatory body in the public sphere that put checks and balances in place and requires accountability could significantly improve what is currently a dysfunctional research misconduct system.

The institutional behavior described in the many examples in this book exemplifies a lack of accountability that permits universities to unabashedly favor their strategic interests, while stepping on the rights of faculty, students and post-docs. Confidentiality has been used to muzzle anyone who reminds them of their ethical responsibility. Although perhaps an egregious example, my experience of the dysfunction of the research misconduct system in the hexagon case study is not an isolated incident. In fact, now that my experience has been reported in numerous journals and newspapers, I hear about incidents on a regular basis at nearly every institution I visit, all over the world. For the most part, these are kept quiet. I must conclude that the documented cases, the peer review crisis, the record retraction rate and lack of reproducibility are but the tip of an iceberg. Perhaps one of the reasons that I have become so ardent about this issue is that prior to being blindsided by the events that led to the hexagon case I was among the skeptics who believed that descriptions of ethical problems were exaggerated. When I stumbled into the morass, I certainly was not looking for falsification or fabrication. I resisted believing it was possible for a period of months. The suggestion of the respondents that I was a vigilante who bullied them is inaccurate. They confused scientific refutation with rhetoric. I believe in transparency because science functions best when communication is open, and when data are available to all. However, during the 9 years of the hexagon case I learned that the majority of ethical issues in research are kept secret, by universities, as well as federal agencies. I now see that it was a mistake to assume that the openness of science would naturally expose its excesses. Scientists collectively must take on the task of cleaning up the literature and discarding discredited publications. No one can do this for us. Universities are not going to lobby for any change that forces them to expose weaknesses in the research misconduct system.

12.3 Will Transparency in Ethics Help or Hurt the Public Perception of Science?

After the Baltimore case, all stakeholders agreed that Congressional hearings on research misconduct should be avoided. The spotlight on scientific malfeasance was damaging for public perceptions, and therefore future funding of science. Perhaps we do not need a spotlight, but under current regulation we do not even have a flashlight. The conflicts of interest of universities shroud unethical conduct. To avoid conflicts of interest, scientists need an independent authority, but one that is maintained by scientists and is founded on a scientific search for the truth. Is this also an example of the fox guarding the henhouse? If implemented correctly then this is the least corruptible system. Scientists are best positioned to distinguish proper science from fraud or falsification. The scientists who make these judgments should not have any relationship with the scientists they are judging. The fact that they are scientists is not likely to make them more favorably disposed towards fellow scientists. Meaningful confidentiality would be implemented, not to muzzle anyone, but to protect both informants and subjects, precisely as described in the regulation. Violation of confidentiality would be considered retaliation and would automatically involve federal authorities at a different level. I endured years of public slander and libel as the respondents described me as a deranged and jealous individual who had a vendetta. Instead of supporting me, University 1 initiated research misconduct proceedings against me based on allegations that they later admitted had "retaliatory animus."

The fact remains that scientists are deeply disturbed by sloppy science. Scientists in an independent body, as proposed here, would not stand to gain or lose by the judgments they make. Nor would any institution or university. This proposal is fundamentally different from the current system where university administrators are acutely aware of the funding status of any scientists who answer to allegations. A highly funded scientist will likely be exonerated and the whistleblower punished as a matter of course. The institution must protect its funding. The loss of funding in the event of a finding of misconduct can be calculated almost to the dollar. This is an extremely unhealthy climate for judging the veracity of statements or making decisions about which evidence to sequester or how to conduct the inquiry. The today's universities have the power to decide nearly every aspect of a case and to deflect scrutiny in many ways.

Some may question whether frank discussion of the growing ethical problems will damage the scientific community by turning the public against science. I too used to worry about this. It is too late. Many people already believe that scientists are biased because they are paid by companies, non-governmental organizations or political think tanks. The sooner scientists can demonstrate that there is a robust way to adjudicate false statements, the sooner they will regain the public trust. Therefore, it would strengthen science to have a structure in which scientists, and not university administrators, manage ethical issues. University administrators are put in an untenable situation by the current regulations. They are given bad choices and they choose the interests of their institutions since that is how they define their responsibilities. Creating a solution to this problem requires new structures and some measure of financial support to permit scientists to take the time to look into the matter from a disinterested point of view. Negative public perception will only be augmented if we continue on the current path and refuse to acknowledge a growing problem that has been amply documented [7, 8, 446].

If the "fake news" crisis becomes a "fake science" crisis, we will have reached a point of extreme danger for science, but also the ecosystem, economy, and society as a whole. The first step towards a rational approach is to eliminate the sources of absurd delays and violations of basic principles of fairness. Currently implemented confidentiality regulations make this impossible. Scientists need to understand the problem, study it and find solutions, on scientific terms that are not subject to either the economic pressures universities or political and administrative pressures of the government. This is why I believe that a separate body staffed by scientists is the best solution.

12.4 Band Aids: Stricter Peer Review and Tougher Journal Standards

It is obvious that increased individual responsibility for adherence to ethics standards would be aided by more rigorous educational requirements for scientific ethics, stricter rules for peer review and enforcement of rigorous journal standards. Increased training and emphasis on individual responsibility are worthwhile steps but they are unlikely to stem the tide of falsification and fabrication. Many scientists fear that to question the research misconduct system, or to institute new transparency requirements, or ethical standards risks an additional regulatory burden. No scientist wants regulators and members of the public entering laboratories with a mandate to scrutinize the inner workings in the way the Environmental Health and Safety can now do in the arena of waste disposal and safety plans. How do we accomplish the goal of maintaining ethics in science without creating an oppressive bureaucracy?

A greater educational emphasis on ethics in science, with a healthy dose of history and even philosophy of science, would help scientists to put their profession in perspective. In order to achieve any kind of change, scientists will need to recognize that there is a problem and then make an effort to understand the current regulations and their effects. Since a minority of scientists have direct experience with falsification individual recognition of this problem is unlikely to gain traction. For every scientist adversely affected numerous others will have no understanding for the reality of the problem or may even actively disparage a person who has come forward to report ethical violations.

We need to instill more of a group consciousness in the research ethics arena. Scientific research is almost always done by a research group. There needs to be a group culture and sense of a group responsibility. Unless a PI can prove that he or she was misled by a student, the responsibility for image manipulations and other falsifications belongs to *both* the PI and student. In cases where the PI knowingly misdirects a student or training is substandard, the PI is primarily responsible for the infraction. Publication is a group effort and all authors should be aware of what is being presented in a manuscript before it goes to print.

Within the past few years, journals have begun to require confirmation and consent of each author listed in a publication. While it is not fair to assume that the PI is always responsible, it is unconscionable to blame students for the entire matter unless one can prove that the student intentionally deceived other group members. There are many shades of gray in the palette of transgressions of research protocol. It is relatively easy to imagine how certain misrepresentations end up in print because of miscommunication within the group. That is an honest mistake. It is embarrassing, perhaps, but it is not research misconduct, and it should not require an investigation. Research groups need to take responsibility and retract flawed papers on their own, but the independent body proposed here would enforce the retraction in cases where there is an established error of sufficient significance or any type of manipulation, regardless of intent. Sometimes, even an honest mistake ends up becoming a deception when people do not accept responsibility.

A collective integrity should apply to faculty members in a department. The professors should look out for each other and help to avoid ethical problems by enforcing standards in education by admonishing sloppy work in students. While it is difficult to confront a faculty member with their failings, an active department keeps everyone honest by confronting the student in routine presentation or preliminary (pre-candidacy) examinations. One hopes that serious problems will be identified before the final PhD defense, but even at that stage, a message can be sent to research advisor. This is one avenue in the current system to confront breaches of ethics when they occur, but it requires more active engagement and willingness to ask the tough questions. We can hope that education would change the research culture and encourage pro-active engagement by research group members and faculty in a department. But individual experiences of research will tend to lead many scientists to take a path of less resistance, not to confront issues in student presentation and not to delve more deeply to see whether there is evidence of an ethical problem. Up to the present, the individual mandate to act ethically and speak out to correct ethical lapses has failed.

The peer review process has broken down in numerous ways. Reviewers may be too busy, or it may be difficult to find reviewers with appropriate expertise for interdisciplinary projects. Entrenched reviewers may look out for their own interests or fail to appreciate new ideas and new investigators. Another failing of peer review arises from aberrations such as collusion or bias. In extreme cases, there is fake peer review, which has recently surfaced as an epidemic with thousands of publications enshrined in the public domain because of bogus peer review at lax journals. Despite scientists' frustration with the peer review system, it is hard to conceive of a system that has greater fairness or consideration for a wide range of viewpoints in principle. In practice, many of these issues would be ameliorated if peer review were elevated in stature by various means to reward individuals more for their contributions. Recording peer reviews on journal servers has helped to a certain extent. A reward system based on credit for having conducted reviews has value because researchers are usually required to report their annual activities as part of a performance review. Today the reporting of reviewing is done on the honor system and since there is little documentation it is also not given a great deal of weight in an evaluation of research accomplishments. The peer review process would be greatly improved if all journals moved to a double-blind peer review system. Post-publication peer review sanctioned by journals and universities would be another means to encourage more responsible publication. Financial incentives for peer review are likely to introduce perverse incentives and are not recommended.

Journal standards are clearly important in preventing the subculture of misrepresentation from gaining ground. Despite the supposedly higher standards of premier journals, it has frequently been easier to publish a falsified account in a high impact journal than a low impact journal. While this tendency is difficult to quantify, the retraction rate in the luxury journals is quite high. The requirement for evidence is actually lower in the higher impact journals than the subject journals. We can consider the standards in three different journals, a specialty journal in a sub-field, *Journal of Organic Chemistry (JOC)*, a high-profile journal covering a subject wholistically, *Journal of the American Chemical Society (JACS)*, and the high-prestige magazine, *Science*, which tackles newsworthy contributions to all of – as the name implies – science. A publication in the *JOC* must provide exacting analytical evidence that a molecule has been synthesized. The list of required experiments includes mass spectrometry, elemental analysis and nuclear magnetic resonance. Yet the same paper submitted to *JACS* does not face such an exhaustive checklist. A paper can be published with significantly fewer supporting data. The apparent logic is that the more impactful the result, the less background information we need in the initial report. The requirement for supporting evidence has historically been even lower in *Science*. This publishing concept is built on trust. The clear expectation is that a high impact publication with relatively little detail will be quickly followed with other publications in the lower tier journals that flesh out the idea and provide all of the details. While this may still be a useful format, the background information should be required to evaluate the validity of the claims in the high impact journals. Many scientists have used the more lax standards of the high impact journals to publish there and then fail to follow up with appropriately detailed accounts that would permit scrutiny of the claims. The system is distorted such that when a follow-up publication appears in a lower tier journal, the prestige of prior publication can be used to obfuscate conditions or discrepancies. The hexagon case is an excellent example of this strategy. This strategy has been pursued widely in recent years. There is no indication that it is slowing down.

The initial goal of any enforcement action should be to correct the scientific literature. The most desirable solution to ethical problems should be education rather than punitive measures. The goal is to encourage accurate reporting. Courses on ethics are useful, but changes in behavior will require a more concerted effort. The procedures for retraction should be streamlined, but the stigma associated with retraction should simultaneously be mitigated. Here too, it is unlikely that journals by themselves will be able to implement more rigorous enforcement of ethical standards or to change attitudes. There are legal and financial pressures that work against this. The individual approaches currently being implemented to enforce research ethics are band aids. There is no obvious set of incentives at the level of the journals and peer groups to modulate human behavior to achieve the desired outcome. This is why I am proposing a more radical solution to the problem. Indeed, creating yet another bureaucracy may sound onerous, but I believe that scientific ethics need the oversight of an independent institution with nation-wide scope and independence from both the government and universities. The current state of affairs is unacceptable.

12.5 Proposal for a National Research Ethics Oversight Council (NREOC)

The federal government could play a constructive role by mandating that all universities must collectively create a National Research Ethics Oversight Council (NREOC) that adjudicates research misconduct cases. The NREOC would then set standards that are fair for students as well as faculty. It could be funded by a requirement that 0.25% of all federal funding received by universities be placed into a fund for the NREOC. This is not a call for greater federal regulation. It is a suggestion that a government mandate could create an independent institution, which will set a standard for scientists that they collectively could agree to. Scientists would set the standards, term limits of adjudicators, requirements for participation and standards. Collectively scientists would also have the right to question the standards and policies of such a body.

Federal regulation would still have a role to play in the various federal agencies and their OIGs. However, the NREOC would be an independent body that would have paid positions for those who adjudicate research misconduct cases and refer only the most serious cases to the OIG. This would free the OIG from burdensome caseloads and permit a timely adjudication by all institutions. The OIGs have so few agents that they need to focus on the egregious cases. Any case that is not adequately addressed by university investigation but is sufficiently complicated that it will require resources of the OIG will take many years to adjudicate. With the NREOC, adjudication would be conducted on a scientific basis and expertise would be a required component of any evaluation. Unlike the federal agencies, the funding aspect would not be the primary consideration of the NREOC. The first question asked would not be whether the taxpayer was defrauded (as it is now), but rather whether the published work stands up to scrutiny. Greater transparency in deliberations would not weaken science. On the contrary, an expedited scientific process by a recognized authority could protect scientists under attack because of the inconvenient nature of their findings for an investor. If we accept the premise that science is under attack by certain forces with social, political, or economic interests [91], then a central authority composed of scientists would be the best defense against the "fake news" phenomenon that has spread to science in controversial domains such as climate science and environmental protection, and has distorted evidence based medical and public health science in the age of the coronavirus. The NREOC could make recommendations for peer review and other evaluations to minimize the incidence of falsification in articles and grants. Education would be an important mission of the Council as well. While this proposal could eliminate the most glaring aspects of the current dysfunction, it is hoped that the increased awareness it would foster would have an impact at all levels including individual scientists, peer reviewers, and journal editors.

The funding agency OIGs and ORI would rely on the NREOC to do the initial screening of cases to ascertain which ones require full investigation and adjudication. Thus, the OIGs of the various agencies could focus on the serious cases. Given the current crushing case load, it is unlikely that this innovation would change their role of the OIGs, but hopefully it would reduce the case load so that they could effectively and speedily adjudicate the serious cases. The punishment at the federal level (debarment and exclusion from serving as a reviewer for 3 years) would be reserved for serious cases. For the rest of the ethical violations, the punishment would be a swift retraction. Such a system, if implemented as a rapid, fair and accurate post-publication review, would force scientists to rethink rushing to publish a questionable result. Knowing that one's paper could quickly be retracted by a binding committee vote in the NREOC would be a much greater deterrent to self-deception. The problem with the current regulation is not the definitions of the infractions, but it is the conflict of interest and massive inefficiency built into the adjudication process. These issues are so severe that they force one to contemplate alternatives like the NREOC, albeit a potentially new bureaucracy. At least, the NREOC would not belong to any university or to the government and would answer to the scientific community in a more direct way than any current institution. There is an obvious role for significant education in research ethics. Education and a demonstrated commitment to understanding of the legal and ethical issues would be as important as scientific expertise as a qualification for serving in the NREOC.

The responsibilities of scientists on the NREOC would include evaluating whether or not allegations concern honest disagreements or simple errors. The openness of the system would provide authors with feedback in borderline cases, where mistakes have been made but there is no misconduct. Only when scientific principles are introduced into the adjudication process can we hope to reverse the trend that has led to the current situation. Transparency would help all scientists because publication of scientific results is itself an inherently public act.

12.6 How Can We Deflate the Science Bubble?

A science bubble has formed resulting from economic pressure that has driven both individual scientists and universities to push the marketing of science to a degree that can compromise research ethics. For the most part, it is not outright conscious fraud that has created the science bubble. Instead, misrepresentation of science has arisen because of a combination of poor communication, wishful thinking, and sloppy record keeping. Once authors have published a result that has questionable validity they will often struggle to avoid retraction, treading further down a path the ultimately leads to dishonesty. The exaggeration of scientific results that arises in the race for novelty, impact and significance expands the bubble. The interpretation of novelty as revolutionary and paradigm-shifting research has driven scientists into territory where there is a great potential for ethics violations. Science policy should foster the greatest possible openness, data sharing, and discussion to catch unsupported observations at the earliest possible stage. The actions of restricting information, claiming areas of science as one's private domain, and conducting experiments that others cannot check increase the possibility that mistakes will not be discovered. This approach violates the most basic tenant that science must be falsifiable. When inscrutable science proliferates, it resembles the bundling of financial instruments into large mortgage backed securities that caused the U.S. economy to crash in 2008. Investors did not have enough information to know what they were buying. If the increased emphasis on patentable technology results in a culture of secrecy and deniability, society runs the risk of wasted resources and poor investment due to flawed reporting of science. We need to reform the research misconduct system so that it is responsive to the needs of science, rather than for the benefit university administrators who most often merely wish that such problems would vanish. Putting those who manage a large financial institution, a university, directly in charge of adjudicating the financial and scientific malfeasance of their own faculty is a disastrous policy that has led us down a path to science bubbles. Whole areas of science have become beholden to certain research groups, funding agencies and companies in a way that has distorted both the educational and ethical mission of the university. As the waste of public investment becomes evident, we have witnessed an increase in attacks on science and calls for major cuts to scientific funding. The continued existence of this type of corruption within the scientific community feeds into a public backlash against science, which is already well underway. There is an urgent need for ethical publishing and promotion models that return science to the original goal of a search for the truth.

Appendixes

Appendix A: Chronology of Events

2004, *08 NSF funding* received by two professors from University 1.
2005, *01 W.M.Keck funding* received by two professors from University 1 and Franzen.
2005, 05 First of two professors from University 1 moves to University 2.
2005, *07 DoE funding* received by two professors from University 1.
2005, *12 Internal report* in W.M. Keck Center reveals apparent error. Hexagons are mostly carbon.
2006, *03 NSF funding* received by Dr. Franzen.
2006, *11 Inquiry begins at University 1.*
2007, 06 Second of two professors from University 1 moves to University 2.
2007, *07 Investigation begins University 1.* The investigation is the second phase of the process triggered by the findings of the inquiry report that the matter is sufficiently serious to warrant a full examination.
2008, *07 University 1 Investigation Committee* finds falsification of data, deviation from normal research practices, but is split on the decision of research misconduct.
2008, *09 NSF OIG* does not accept the University 1 report as final and decides to investigate.
2008, *11 University 2 Inquiry* concludes that the University 1 investigation committee report is an exoneration and declares that the allegations by Dr. Franzen are "without substance".
2013, *09 NSF OIG* finds that the two professors from University 1 (currently at University 2) committed research misconduct in the Semi-Annual Report to Congress.

S. Franzen, *University Responsibility for the Adjudication of Research Misconduct*, https://doi.org/10.1007/978-3-030-68063-3

2015, *09 NSF Deputy Director issues the final report* and revises the finding against the two professors from University 1. The final report states that the two professors from University 1 falsified data and cites "failure to report to the NSF Director" as grounds for debarment for life. The final report does not find research misconduct, but the letter of reprimand states that the "failure to report" is a final decision that cannot be appealed. Research misconduct can be appealed in court and the respondents have stated publicly that they will appeal any decision of research misconduct.

2016, *02 The Editor-in-Chief of Science, Dr. Marcia McNutt* unilaterally retracts the 2004 Science paper of the two professors from University 1 over their objections.

Appendix B: List of Abbreviations

COI	Conflict of Interest
DoD	Department of Defense
DOE	Department of Energy
HHS	Health and Human Services
IC	Investigation Committee
JACS	Journal of the American Chemical Society
JMC	Journal of Materials Chemistry
JPC	Journal of Physical Chemistry
JCE	Journal of Chemical Education
NIH	National Institutes of Health
NNI	National Nanotechnology Initiative
NSF	National Science Foundation
OIG	Office of the Inspector General
ORI	Office of Research Integrity
PNAS	Proceedings of the National Academy of Science
RIO	Research Integrity Officer
SST	The Committee on Space, Science and Technology of the U.S. Congress
TTO	Technology Transfer Office
University 1	University 1
University 2	University 2
USDA	United States Department of Agriculture

References

1. Steen, R.G., A. Casadevall, and F.C. Fang. 2013. Why has the Number of Scientific Retractions Increased? *Plos One* 8.
2. Greenberg, D. 2007. *Science for Sale: The Perils, Rewards, and Delusions of Campus Capitalism*. Chicago: University of Chicago Press.
3. Washington University. 2015. University of Washington Provost Town Hall https://www.youtube.com/watch?v=F7GkKIs8N5o. 24 February 2015, minute 14:50.
4. Washburn, J. University Inc. 2006. *The Corporate Corruption of Higher Education*. New York: Basic Books.
5. Edwards, M.A., and S. Roy. 2017. Academic Research in the 21st Century: Maintaining Scientific Integrity in a Climate of Perverse Incentives and Hypercompetition. *Environmental Engineering Science* 34: 51–61.
6. Kaiser, J. 2014. Top U.S. Scientific Misconduct Officer Quits in Frustration with Bureaucracy. *Science*, March 12.
7. Judson, H.F. 2004. *The Great Betrayal: Fraud in Science*. Orlando: Harcourt Press.
8. Broad, W., and N. Wade. 1983. *Betrayers of the Truth: Fraud and Deceit in the Halls of Science*. London: Century Publishing.
9. Reich, E. 2009. *Plastic Fantastic*. New York: Palgrave Macmillan.
10. Kelves, D.J. 1998. *The Baltimore Case: A Trial of Politics, Science, and Character*. New York: W.W. Norton.
11. Resnik, D.B., J.L. Ariansen, J. Jamal, and G.E. Kissling. 2016. Institutional Conflict of Interest Policies at US Academic Research Institutions. *Academic Medicine* 91: 242–246.
12. CFR689. Code of Federal Regulations 689. https://www.gpo.gov/fdsys/pkg/CFR-2015-title45-vol3/xml/CFR-2015-title45-vol3-part689.xml.
13. Crocker, J. 2011. The Road to Fraud Starts with a Single Step. *Nature* 479: 151.
14. Oransky, I. 2014. Scientists, Do You Feel Bullied by Critics? These Chemists Do. *Retraction Watch*. http://retractionwatch.com/2014/2004/2002/scientists-do-you-feel-bullied-these-two-chemists-do/. April 2.
15. Alford, C.F. 2001. *Whistleblowers: Broken Lives and Organizational Power*. Ithaca: Cornell University Press.
16. Jacquet, L. 2005. Director, March of the Penguins, Wild Bunch, National Geographic Films, and Bonne Poche.
17. Carson, R. 2002. *Silent Spring*, 13. New York: Houghton Mifflin. first pub. 1962.
18. Lancaster, C. 2015. The Acid Test for Biological Science: STAP Cells, Trust, and Replication. *Science and Engineering Ethics* 22: 147–167.
19. Hawkes, N. 2014. Investigation into Stem Cell Research Claims Finds Scientific Misconduct by Researcher and Lack of Supervision by Coauthors. *British Medical Journal* 348: g2563.

S. Franzen, *University Responsibility for the Adjudication of Research Misconduct*, https://doi.org/10.1007/978-3-030-68063-3

20. Normile, D. 2014. Stem Cell Research RIKEN Panel Finds Misconduct in Controversial Paper. *Science* 344: 23–23.
21. Carpenter, S. 2012. Harvard Psychology Researcher Committed Fraud, U.S. Investigation Concludes. *ScienceInsider*, September 6.
22. Reich, E.S. 2007. Disputed Inquiry Clears Bubble-Fusion Engineer. *Nature* 445: 690–691.
23. Chung, S., et al. 2013. The Formation of Pd Nanocrystals from $Pd_2(dba)_3$ Microcrystals. *Particle & Particle Systems Characterization* 30: 280–286.
24. Franzen, S., M. Cerruti, D.N. Leonard, and G. Duscher. 2007. The Role of Selection Pressure in RNA-Mediated Evolutionary Materials Synthesis. *Journal of the American Chemical Society* 129: 15340–15346.
25. Franzen, S., and D.N. Leonard. 2011. Analysis of RNA-Mediated Materials Synthesis Using Magnetic Selection. *Journal of Physical Chemistry C* 115: 9335–9343.
26. Leonard, D.N., M. Cerruti, G. Duscher, and S. Franzen. 2008. Interfacial and Solvent Effects Govern the Formation of tris(dibenzylidenacetone)dipalladium(0) Microstructures. *Langmuir* 24: 7803–7809.
27. Leonard, D.N., G. Duscher, and S. Franzen. 2008. Eletter Concerning "RNA-Mediated Metal-Metal Bond Formation in the Synthesis of Hexagonal Palladium Nanoparticles". *Science e-Letter*.
28. Leonard, D.N., and S. Franzen. 2009. Is Pd-2(DBA)(3) a Feasible Precursor for the Synthesis of Pd Nanoparticles? *Journal of Physical Chemistry C* 113: 12706–12714.
29. Gugliotti, L.A., D.L. Feldheim, and B.E. Eaton. 2004. RNA-Mediated Metal-Metal Bond Formation in the Synthesis of Hexagonal Palladium Nanoparticles. *Science* 304: 850–852.
30. Neff, J. 2014. In notebook at NCSU, a 'smoking gun'. *News and Observer*, January 20.
31. Eaton, B., D. Feldheim, M. Dolska, and L. Gugliotti. 2003. Novel Methods of Inorganic Compound Discovery and Synthesis U.S. Patent Application US 20050136439 A1.
32. Neff, J. 2014. NC State Professor Uncovers Problems in Lab Journal. *New and Observer*, January 19.
33. Neff, J. 2014. Congressmen Push NCSU on Case of Flawed Research. *Raleigh News and Observer*, July 17.
34. Neff. J. 2016. Former NCSU Scientists Reprimanded, Lose Future Funding over 'Misleading' Research. *News and Observer*, January 8.
35. Kuta, S. 2016. National Science Foundation Reprimands CU-Boulder Prof over Research Practices. *Daily Camera*, November 1.
36. Borman, S. 2016. Nanoparticle Synthesis Paper Retracted After 12 Years. *Chemical & Engineering News* 94: 37–38.
37. Mervis, J. 2016. NSF Breaks New Ground in Reprimanding Authors of Flawed Science Paper. *ScienceInsider*, February 4.
38. Kuta, S. 2014. CU-Boulder Scientists Speak Out on Research Misconduct Claim. *Boulder Daily Camera*, February 7.
39. Franzen, S., and Leonard, D.N.A. 2010. Critical Assessment of RNA-Mediated Materials Synthesis. *Material Research Society Symposium Proceedings* 1271: 1272-PP1204-1203.
40. Franzen, S. 2011. Determination of the Solubility Limit of Tris(dibenzylideneacetone) dipalladium(0) in Tetrahydrofuran/Water Mixtures. *Journal of Chemical Education* 88: 619–623.
41. Reich, E.S. 2011. Acrimony over Nanoconstruction. *Nature*, August 23.
42. N.S. Foundation. 2015. https://www.nsf.gov/oig/case-closeout/A06110054.pdf. A06110054.
43. Gugliotti, L.A. 2006. North Carolina State University.
44. Gugliotti, L.A., D.L. Feldheim, and B.E. Eaton. 2005. RNA-Mediated Control of Metal Nanoparticle Shape. *Journal of American Chemical Society* 127: 17814–17818.
45. Liu, D.G., et al. 2006. RNA-Mediated Synthesis of Palladium Nanoparticles on Au Surfaces. *Langmuir* 22: 5862–5866.
46. Franzen, S. 2013. Comment on 'Cooperativity Between Two Selected RNA Pdases in the Synthesis of Pd Nanoparticles' by J. L. Rouge et al. *Journal of Materials Chemistry* 2010, 20: 8394–8398. *Journal of Materials Chemistry B* 1: 6339–6341.

47. Tuerk, C., and L. Gold. 1990. Systematic Evolution of Ligands by Exponential Enrichment – RNA Ligands to Bacteriophage-T4 DNA-Polymerase. *Science* 249: 505–510.
48. Ellington, A.D., and J.W. Szostak. 1990. In vitro Selection of RNA Molecules that Bind Specific Ligands. *Nature* 346: 818–822.
49. Gold, L. 2004. Photoaptamer-Based High-Density Proteomic Arrays. *Molecular & Cellular Proteomics* 3: S2–S2.
50. Gold, L., et al. 2010. Aptamer-Based Multiplexed Proteomic Technology for Biomarker Discovery. *Plos One* 5.
51. Gold, L. et al. 2010. Aptamers and the RNA World, Past and Present. *Cold Spring Harbor Perspectives in Biology*. https://doi.org/10.1101/cshperspect.a003582.
52. Golden, M.C., B.D. Collins, M.C. Willis, and T.H. Koch. 2000. Diagnostic Potential of PhotoSELEX-Evolved ssDNA Aptamers. *Journal of Biotechnology* 81: 167–178.
53. Agresti, J.J., B.T. Kelly, A. Jaschke, and A.D. Griffiths. 2005. Selection of Ribozymes that Catalyse Multiple-Turnover Diels-Alder Cycloadditions by Using in vitro Compartmentalization. *Proceedings of the National Academy of Sciences of the United States America* 102: 16170–16175.
54. Furtig, B., et al. 2007. Time-Resolved NMR Studies of RNA Folding. *Biopolymers* 86: 360–383.
55. Keiper, S., D. Bebenroth, B. Seelig, E. Westhof, and A. Jaschke. 2004. Architecture of a Diels-Alderase Ribozyme with a Preformed Catalytic Pocket. *Chemical Biology* 11: 1217–1227.
56. Kisseleva, N., S. Kraut, A. Jaschke, and O. Schiemann. 2007. Characterizing Multiple Metal Ion Binding Sites Within a Ribozyme by Cadmium-Induced EPR Silencing. *Hfsp Journal* 1: 127–136.
57. Manoharan, V., B. Furtig, A. Jaschke, and H. Schwalbe. 2009. Metal-Induced Folding of Diels-Alderase Ribozymes Studied by Static and Time-Resolved NMR Spectroscopy. *Journal of American Chemical Society* 131: 6261–6270.
58. Jia, T.Z., A.C. Fahrenbach, N.P. Kamat, K.P. Adamala, and J.W. Szostak. 2016. Oligoarginine Peptides Slow Strand Annealing and Assist Non-enzymatic RNA Replication. *Nature Chemistry* 8: 915–921.
59. Jia, T.Z. et al. 2017. Oligoarginine Peptides Slow Strand Annealing and Assist Non-enzymatic RNA Replication (Retraction of Vol 8, Pg 915, 2016). *Nature Chemistry* 9: 1286–1286.
60. Rybarczyk, A., et al. 2015. New in Silico Approach to Assessing RNA Secondary Structures with Non-canonical Base Pairs. *BMC Bioinformatics* 16: 276.
61. Cech, T.R. 2012. The RNA Worlds in Context. *Cold Spring Harbor Perspectives in Biology* 4.
62. Kruger, K., et al. 1982. Self-Splicing RNA – Auto-Excision and Auto-Cyclization of the Ribosomal-RNA Intervening Sequence of Tetrahymena. *Cell* 31: 147–157.
63. Joyce, G.F. 1989. Amplification, Mutation and Selection of Catalytic RNA. *Gene* 82: 83–87.
64. Cech, T.R. 2013. How a Chemist Looks at RNA. *Angewandte Chemie-International Edition* 52: 75–78.
65. Chung, S.W., et al. 2008. Scanning Probe-Based Fabrication of 3D Nanostructures via Affinity Templates, Functional RNA, and Meniscus-Mediated Surface Remodeling. *Scanning* 30: 159–171.
66. Jolly, J. 2017. Why I Failed to Catch Canada's Worst Serial Killer. *BBC World Service*, June 1.
67. Adamiak, R.W., and P. Gornicki. 1985. Hypermodified Nucleosides of Transfer-RNA – Synthesis, Chemistry and Structural Features of Biological Interest. *Progress in Nucleic Acid Research and Molecular Biology* 32: 27–74.
68. Antczak, M., et al. 2016. New functionality of RNAComposer: An Application to Shape the Axis of miR160 Precursor Structure. *Acta Biochimica Polonica* 63: 737–744.
69. Biesiada, M., K. Pachulska-Wieczorek, R.W. Adamiak, and K.J. Purzycka. 2016. RNAComposer and RNA 3D Structure Prediction for Nanotechnology. *Methods* 103: 120–127.
70. Cieslak, M., J. Szymanski, R.W. Adamiak, and C.S. Cierniewski. 2003. Structural Rearrangements of the 10-23 DNAzyme to Beta 3 Integrin Subunit mRNA Induced by

Cations and Their Relations to the Catalytic Activity. *Journal of Biological Chemistry* 278: 47987–47996.
71. Heydenreich, A., et al. 1993. The Complex Between Ribonuclease T1 and 3′GMP Suggests Geometry of Enzymatic Reaction Path – An X-ray Study. *European Journal of Biochemistry* 218: 1005–1012.
72. Lukasiak, P., et al. 2015. RNAssess-a Web Server for Quality Assessment of RNA 3D Structures. *Nucleic Acids Research* 43: W502–W506.
73. Olejniczak, M., et al. 2002. The bulge region of HIV-1 TAR RNA Binds Metal Ions in Solution. *Nucleic Acids Research* 30: 4241–4249.
74. Pachulska-Wieczorek, K., L. Blaszczyk, M. Biesiada, R.W. Adamiak, and K.J. Purzycka. 2016. The Matrix Domain Contributes to the Nucleic Acid Chaperone Activity of HIV-2 Gag. *Retrovirology* 13: 18.
75. Popenda, M., et al. 2012. Automated 3D Structure Composition for Large RNAs. *Nucleic Acids Research* 40: e112.
76. Purzycka, K.J., et al. 2015. Automated 3D RNA Structure Prediction Using the RNAComposer Method for Riboswitches. In *Computational Methods for Understanding Riboswitches*, ed. S.J. Chen and D.H. Burke Aguero, vol. 553, 3–34. Amsterdam: Elsevier.
77. Popenda, M., J. Milecki, and R.W. Adamiak. 2004. High Salt Solution Structure of a Left-Handed RNA Double Helix. *Nucleic Acids Research* 32: 4044–4054.
78. Harvey, P.D., F. Adar, and H.B. Gray. 1989. Spectroscopic Properties of Binuclear Palladium(0) and Platinum(0) Dibenzylideneacetone Complexes. *Journal of American Chemical Society* 111: 1312–1315.
79. Gugliotti, L.A., D.L. Feldheim, and B.E. Eaton. 2008. Eletter Concerning "RNA-Mediated Metal-Metal Bond Formation in the Synthesis of Hexagonal Palladium Nanoparticles". *Science Eletter*.
80. Gugliotti, L.A., D.L. Feldheim and B.E. Eaton 2009. RNA-Mediated Control of Metal Nanoparticle Shape (vol 127, pg 17814, 2005). *Journal of the American Chemical Society* 131: 11634–11634.
81. Rouge, J.L., C.J. Ackerson, D.L. Feldheim, and B.E. Eaton. 2010. Cooperativity Between Two Selected RNA Pdases in the Synthesis of Pd Nanoparticles. *Journal of Materials Chemistry* 20: 8394–8398.
82. Collins, H., and T. Pinch. 1993. *The Golem: What Everyone Should Know About Science*. Cambridge: *Cambridge University Press*.
83. Kolata, G. 2018. Harvard Calls for Retraction of Dozens of Studies by Noted Cardiac Researcher. *The New York Times*, October 15.
84. Kolata, G. 2018. He Promised to Restore Damaged Hearts. Harvard Says His Lab Fabricated Research. *New York Times* October 29.
85. Markman, A. 2010. Why Science is Self-Correcting. *Psychology Today*, August 10.
86. Estes, S. 2012. The Myth of Self-Correcting Science. *The Atlantic*, December 20.
87. Hettinger, T.P. 2010. Misconduct: Don't Assume Science is Self-Correcting. *Nature* 466: 1040–1040.
88. F.I.B. Research. 1981. Hearings before the Subcommittee on Investigations and Oversight of the Committee on Science and Technology, U.S. House of Representatives, 97th Congress. U.S. Government Printing Office, No. 77-661, Washington, 1981 March 31–April 1, 1981, 65–66.
89. Mann, M.E., R.S. Bradley, and M.K. Hughes. 1998. Global-Scale Temperature Patterns and Climate Forcing over The Past Six Centuries. *Nature* 392: 779–787.
90. Briffa, K.R., et al. 1998. Trees Tell of Past Climates: But Are They Speaking Less Clearly Today? *Philosophical Transactions of the Royal Society of London Series B-Biological Sciences* 353: 65–73.
91. Otto, S. 2016. The War on Science. *Milkweed Editions*.
92. Sheppard, N. 2011. ClimateGate 2.0: 5,000 New E-mails Confirm Pattern of Deception and Collusion by Alarmists. *FOX News*, November 23.

93. AAAS. 2011. Statement of the Board of Directors of the American Association for the Advancement of Science Regarding Personal Attacks on Climate Scientists, June 28.
94. Redman, B.K. 2013. *Research Misconduct Policy in Biomedicine: Beyond the Bad-Apple Approach*. Cambridge, MA: The MIT Press.
95. Calabrese, E.J. 2018. Muller's Nobel Prize Research and Peer Review. *Philosophy Ethics and Humanities in Medicine* 13: 15.
96. Colquhoun, D. 2011. Publish-or-Perish: Peer Review and the Corruption of Science. *Guardian*, September 5.
97. Lacetera, N., and L. Zirulia. 2011. The Economics of Scientific Misconduct. *Journal of Law Economics & Organization* 27: 568–603.
98. Qiu, J. 2010. Publish or Perish in China. *Nature* 463: 142–143.
99. Bornmann, L., I. Nast, and H.D. Daniel. 2008. Do Editors and Referees Look for Signs of Scientific Misconduct When Reviewing Manuscripts? A Quantitative Content Analysis of Studies That Examined Review Criteria and Reasons for Accepting and Rejecting Manuscripts for Publication. *Scientometrics* 77: 415–432.
100. Chawla, D.S. 2017. Fake Peer Review Hits RSC Journals. *Chemistry World*, August 30.
101. Resnik, D.B., L.M. Rasmussen, and G.E. Kissling. 2015. An International Study of Research Misconduct Policies. *Accountability in Research-Policies and Quality Assurance* 22: 249–266.
102. Fanelli, D., and V. Lariviere. 2016. Researchers' Individual Publication Rate has not Increased in a Century. *Plos One* 11: e0149504.
103. Barbash, F. 2015. Major Publisher Retracts 43 Scientific Papers Amid Wider Fake Peer-Review Scandal. *Washington Post*, March 27.
104. Fergusen, C., A. Marcus, and I. Oransky. 2014. Publishing: The Peer-Review Scam. *Nature*, November 24.
105. McCook, A. 2017. A New Record: Major Publisher Retracting More Than 100 Studies from Cancer Journal over Fake Peer Reviews. *Retraction Watch*, April 20.
106. Oransky, I. 2012. South Korean Plant Compound Researcher Faked Email Addresses so He Could Review His Own Studies. *Retraction Watch*, August 24.
107. McCook, A. 2016. Springer, BMC Retracting Nearly 60 Papers for Fake Reviews and Other Issues. *Retraction Watch*, November 1.
108. Frank, E. 1996. Editors' Requests of Peer Reviewers: A Study and a Proposal. *Preventive Medicine* 25: 102–104.
109. Koppers, L., H. Wormer, and K. Ickstadt. 2017. Towards a Systematic Screening Tool for Quality Assurance and Semiautomatic Fraud Detection for Images in the Life Sciences. *Science and Engineering Ethics* 23: 1113–1128.
110. Goldner, J.A. 1998. The Unending Saga of Legal Controls Over Scientific Misconduct: A Clash of Cultures Needing Resolution. *American Journal of Law & Medicine* 24: 293–343.
111. Ferguson, C. 2014. University of Utah Investigation Fingers Chem Engineering Grad Student for Misconduct. *Retraction Watch*, December 6.
112. Feldman, M., I. Feller, J. Bercovitz, and R. Burton. 2002. Equity and the Technology Transfer Strategies of American Research Universities. *Management Science* 48: 105–121.
113. Bonito, A.J., S.L. Titus, and D.E. Wright. 2012. Assessing the Preparedness of Research Integrity Officers (RIOs) to Appropriately Handle Possible Research Misconduct Cases. *Science and Engineering Ethics* 18: 605–619.
114. Loui, M.C. 2002. Seven Ways to Plagiarize: Handling Real Allegations of Research Misconduct. *Science and Engineering Ethics* 8: 529–539.
115. Sullivan, A. 2009. Cash-Strapped State Schools Being Forced to Privatize. *Time*, April 23.
116. Randi, J. 1988. The Detection of Fraud and Fakery. *Experientia* 44: 287–290.
117. Franklin, A., A.W.F. Edwards, D.J. Fairbanks, D.L. Hartl, and T. Seidenfeld. 2008. *Ending the Mendel-Fischer controversy*. Pittsburgh: University of Pittsburg Press.
118. Fisher, R.A. 1918. The Correlation Between Relatives on the Supposition of Mendelian Inheritance. *Transactions of the Royal Society of Edinburgh* 52: 399–433.

119. Kuhn, T.S. 1962. *The Structure of Scientific Revolutions*. Chicago: University of Chicago Press.
120. Goldberg, D.M. 2011. Science at the Crossroads: Fact or Fiction. *Journal of Medical Biochemistry* 30: 79–92.
121. Eisley, L. 1979. *Darwin and Mysterious Mr. X. E. P.* New York: Dutton.
122. Brown, E.N., and S. Ramaswamy. 2007. Quality of Protein Crystal Structures. *Acta Crystallographica Section D-Biological Crystallography* 63: 941–950.
123. George, S.L. 2016. Research Misconduct and Data Fraud in Clinical Trials: Prevalence and Causal Factors. *International Journal of Clinical Oncology* 21: 15–21.
124. Marcovitch, H. 2007. Misconduct by Researchers and Authors. *Gaceta Sanitaria* 21: 492–499.
125. Quandt, R.E. 2012. Some Models of Academic Corruption. *European Journal of Law and Economics* 34: 63–75.
126. Stein, C. 2015. Scientific fraud. *Trends in Anaesthesia and Critical Care* 5: 76–79.
127. Bowler, P.J. 2003. *Evolution: The History of an Idea*. Berkeley: University of California Press.
128. Barber, B. 1961. Resistance by Scientists to Scientific Discovery- This Source of Resistance has yet to be Given Scrutiny Accounted Religious and Ideological Sources. *Science* 134: 596–602.
129. Gordin, M.D. 2012. How Lysenkoism Became Pseudoscience: Dobzhansky to Velikovsky. *Journal of History of Biology* 45: 443–468.
130. A normal distribution is given by the function $P(x)=\frac{1}{\sqrt{2\pi}\sigma}exp\left\{-\frac{(x-\mu)^2}{2\sigma^2}\right\}$ where μ is the mean value and σ is the standard deviation. This is the function plotted in Figures 4 and 5. In practical applications the standard deviation is replaced by the standard error $\sigma_e = \sigma/\sqrt{N}$, which I have omitted to simplify the explanation.
131. The Bayesian conditional probability is calculated using the equation $P_{conditional}=\frac{n!}{x!(n-x)!}\theta^x(1-\theta)^{n-x}$ where θ is the prior estimate of the probability that a new result will be obtained, x is the number of measurements that are consistent with the new result and n is the total number of measurements.
132. Bayes, F.R.S. 1763. An Essay Towards Solving a Problem in the Doctrine of Chances. Communicated by Mr. Price, in a Letter to John Canton. *Philosophical Transactions of the Royal Society of London* 53: 370–418.
133. Gosset, S.W.S. 1908. The Probable Error of a Mean. *Biometrika* 6: 1–25.
134. Ziliak, S. 2008. Guinnessometrics: The Economic Foundation of Student's t. *Journal of Economic Perspectives* 22: 199–216.
135. Gorard, S. 2018. Damaging Real Lives Through Obstinacy: Re-emphasising Why Significance Testing is Wrong (vol 21, pg 1, 20116). *Sociological Research Online* 23: 285–285.
136. Locascio, J.J. 2017. Results Blind Science Publishing. *Basic and Applied Social Psychology* 39: 239–246.
137. Simmons, J.P., L.D. Nelson, and U. Simonsohn. 2011. False-Positive Psychology: Undisclosed Flexibility in Data Collection and Analysis Allows Presenting Anything as Significant. *Psychological Science* 22: 1359–1366.
138. Mcampo, E.J., and D.R. Lalande. 2012. A Peculiar Prevalence of p Values Just Below .05. *Quarterly Journal of Experimental Psychology* 65: 2271–2279.
139. Gorard, S. 2016. Damaging Real Lives Through Obstinacy: Re-emphasising Why Significance Testing is Wrong. *Sociological Research Online* 21: 102–115.
140. Cumming, G. 2014. The New Statistics: Why and How. *Psychological Science* 25: 7–29.
141. Nicholson, J., and S. McCusker. 2016. Damaging the Case for Improving Social Science Methodology Through Misrepresentation: Re-asserting Confidence in Hypothesis Testing as a Valid Scientific Process. *Sociological Research Online* 21: 136–147.
142. Spreckelsen, T.F., and M. Is van der Horst. 2016. Banning Significance Testing the Best Way to Improve Applied Social Science Research? – Questions on Gorard (2016). *Sociological Research Online* 21: 95–105.

143. Kirk, R.E. 2003. The Importance of Effect Magnitude. In *Handbook of Research Methods in Experimental Psychology*, ed. S.F. Davis, 83–105. Malden, MA: Blackwell.
144. Dienes, Z., and N. McLatchie. 2018. Four Reasons to Prefer Bayesian Analyses over Significance Testing. *Psychonomic Bulletin & Review* 25: 207–218.
145. Howson, C., and P. Urbach. 1989. *Scientific Reasoning: The Bayesian Approach*. La Salle, IL: Open Court.
146. Reich, B.J., and S. Ghosh. 2019. *Bayesian Statistical Methods*. Boca Raton: Chapman and Hall/CRC Press.
147. Kruschke, J.K., and T.M. Liddell. 2018. The Bayesian New Statistics: Hypothesis Testing, Estimation, Meta-analysis, and Power Analysis from a Bayesian Perspective. *Psychonomic Bulletin & Review* 25: 178–206.
148. Earman, J. 1992. *Bayes or Bust? A Critical Account of Bayesian Confirmation Theory*. Cambridge, MA: MIT Press.
149. Popper, K. 1983. *Realism and the Aim of Science: From the Postscript to The Logic of Scientific Discovery*. London/New York: Routledge.
150. Popper, K.R., ed. 2006. *The Logic of Scientific Discovery*. London/New York: Routledge/ Taylor & Francis.
151. Feyerabend, P. 1975. Against Method. *New Left Books*.
152. Kuntz, M. 2012. The Postmodern Assault on Science If All Truths are Equal, Who Cares What Science has to Say? *Embo Reports* 13: 885–889.
153. Sokal, A. 1996. Transgressing the Boundaries: Towards a Transformative Hermeneutics of Quantum Gravity. *Social Text* 46/47: 217–252.
154. Sokal, A. 1996. A Physicist Experiments with Cultural Studies. *Lingua Franca* 6: 62–64.
155. Spelda, D. 2017. Kepler in the Early Historiography of Astronomy (1615-1800). *Journal for the History of Astronomy* 48: 381–404.
156. Russell, J.L. 1964. Kepler's Law of Planetary Motion – 1609-1666. *British Journal for the History of Science* 2: 1–24.
157. de Bruyn Ouboter, R. 1997. Heike Kamerlingh Onnes's Discovery of Superconductivity. *Scientific American* 276: 98–103.
158. Bednorz, J.G., and K.A. Mueller. 1986. Possible High TC Superconductivity in the Ba-La-Cu-O System. *Z. Physik B* 64: 189–193.
159. Waldrop, M.M. 1989. Catalytic RNA Wins Chemistry Nobel. *Science* 246: 325.
160. Cech, T.R. 1990. Nobel Lecture: Self-Splicing and Enzymatic Activity of an Intervening Sequence RNA from Tetrahymena. *Bioscience Reports* 10: 239–261.
161. Cech, T., and F. Steele. 2013. The (Noncoding) RNA World. *Nucleic Acid Therapeutics* 23 (1): 1.
162. Perutz, M.F., J.C. Kendrew, and H.C. Watson. 1965. Structure and Function of Haemoglobin 2. Some Relations Between Polypeptide Chain Configuration and Amino Acid Sequence. *Journal of Molecular Biology* 13: 669–678.
163. Labinger, J.A., and S.J. Weininger. 2004. Controversy in Chemistry: What Counts as Evidence? Two Studies in Molecular Structure. *Angewandte Chemie-International Edition* 43: 2612–2619.
164. Collman, J.P., R.R. Gagne, T.R. Halbert, J.C. Marchon, and C.A. Reed. 1973. Reversible Oxygen Adduct Formation in Ferrous Complexes Derived from a Picket Fence Porphyrin Model for Oxymyoglobin. *Journal of the American Chemical Society* 95: 7868–7870.
165. Ivanov, D., et al. 1994. Determination of CO Orientation in Myoglobin in Single-Crystal Infrared Linear Dichroism. *Journal of the American Chemical Society* 116: 4139–4140.
166. Vojtechovsky, J., K. Chu, J. Berendzen, R.M. Sweet, and I. Schlichting. 1999. Crystal Structures of Myoglobin-Ligand Complexes at Near-Atomic Resolution. *Biophysical Journal* 77: 2153–2174.
167. Slebodnick, C., and J.A. Ibers. 1997. Myoglobin Models and Steric Origins of the Discrimination Between O-2 and CO. *Journal of Biological Inorganic Chemistry* 2: 521–525.
168. Kleywegt, G.J., and T.A. Jones. 1995. Where Freedom is Given, Liberties are Taken. *Structure* 3: 535–540.

169. Kleywegt, G.J. and T.A. Jones 1996. Phi/psi-chology: Ramachandran Revisited. *Structure* 4: 1395–1400.
170. Kleywegt, G.J. and T.A. Jones 1997. Model Building and Refinement Practice. In *Macromolecular Crystallography, Part B*, ed. C.W. Carter and R.M. Sweet, vol. 277, 208–230. San Diego: Academic Press.
171. Rupp, B., A. Wlodawer, W. Minor, J.R. Helliwell, and M. Jaskolski. 2016. Correcting the Record of Structural Publications Requires Joint Effort of the Community and Journal Editors. *The FEBS Journal* 283: 4452–4457.
172. Stanfield, R., E. Pozharski, and B. Rupp. 2016. Comment on Three X-ray Crystal Structure Papers. *Journal of Immunology* 196: 521–524.
173. Weichenberger, C.X., E. Pozharski, and B. Rupp. 2013. Visualizing Ligand Molecules in Twilight Electron Density. *Acta Crystallographica Section F-Structural Biology and Crystallization Communications* 69: 195–200.
174. Weichenberger, C.X., E. Pozharski, and B. Rupp 2017. Twilight Reloaded: The Peptide Experience. *Acta Crystallographica Section D-Structural Biology* 73: 211–222.
175. Minor, W., Z. Dauter, J.R. Helliwell, M. Jaskolski, and A. Wlodawer. 2016. Safeguarding Structural Data Repositories against Bad Apples. *Structure* 24: 216–220.
176. Shabalin, I., Z. Dauter, M. Jaskolski, W. Minor, and A. Wlodawer. 2015. Crystallography and Chemistry Should Always Go Together: A Cautionary Tale of Protein Complexes with Cisplatin and Carboplatin. *Acta Crystallographica Section D-Biological Crystallography* 71: 1965–1979.
177. Weber, M. 2004. *The Vocation Lectures*, Eds. D. Owen and T.B. Strong. Indianapolis/Cambridge: Hackett Publishing.
178. Merton, R.K. 1957. Priorities in Scientific Discovery – A Chapter in the Sociology of Science. *American Sociological Review* 22: 635–659.
179. Lane, N., and T. Kalil. 2005. The National Nanotechnology Initiative: Present at the Creation. *Issues in Science and Technology* 21 (Summer): 49–54.
180. Cole, S., J.R. Cole, and G.A. Simon. 1981. Chance and Consensus in Peer-Review. *Science* 214: 881–886.
181. Cole, J.R., and S. Cole. 1972. Ortega Hypothesis. *Science* 178: 368–375.
182. Lee, P. 2013. Patents and the University. *Duke Law Journal* 63: 1–87.
183. Bud, R. 2008. Upheaval in the Moral Economy of Science? Patenting, Teamwork and the World War II Experience of Penicillin. *History and Technology* 24: 173–190.
184. George, G. 2005. Learning to be Capable: Patenting and Licensing at the Wisconsin Alumni Research Foundation 1925-2002. *Industrial and Corporate Change* 14: 119–151.
185. Rosenberg, N., and R.R. Nelson. 1994. American University and Technical Advance in Industry. *Research Policy* 23: 323–348.
186. Kumar, M.N. 2010. Ethical Conflicts in Commercialization of University Research in the Post-Bayh-Dole Era. *Ethics & Behavior* 20: 324–351.
187. Washington, D.C. 1998. Factors that Influence Facilities and Administrative Cost Rates at Research Intensive Universities. Council on Governmental Relations.
188. Larsen, M.T. 2011. The Implications of Academic Enterprise for Public Science: An Overview of the Empirical Evidence. *Research Policy* 40: 6–19.
189. U.S. Congress. 2007. Federal Support for Research and Development. Congressional Budget Office.
190. Blumenthal, D. 1996. Relationships Between Academic Institutions and Industry in the Life Sciences – An Industry Survey. *The New England Journal of Medicine* 334: 368–373.
191. Berman, E.P. 2008. Why did Universities Start Patenting? Institution-Building and the Road to the Bayh-Dole Act. *Social Studies of Science* 38: 835–871.
192. Valdivia, W.D. 2013. *University Start-Ups: Critical for Improving Technology Transfer*. Brookings Institution.

193. McClure, K.R. 2016. Building the Innovative and Entrepreneurial University: An Institutional Case Study of Administrative Academic Capitalism. *Journal of Higher Education* 87: 516–543.
194. Pflumm, M. 2011. NIH Funding Rates Drop to Record Lows. *Nature Medicine* 17: 637.
195. Hollon, T. 2001. NIH Budget Maintains Doubling Momentum. *The Scientist*, January 22.
196. Report. 2010. Chapter 5. *National Science Board.*
197. Harris, R. 2017. *Rigor Mortis: How Sloppy Science Creates Worthless Cures, Crushes Hope, and Wastes Billions*. New York: Basic Books.
198. Basch, D.L. 1997. Private Colleges' Pricing Experience in the early 1990s: The Impact of Rapidly Increasing College-Funded Grants. *Research in Higher Education* 38: 271–296.
199. Metlay, G. 2006. Reconsidering Renormalization: Stability and Change in 20th-Century Views on University Patents. *Social Studies of Science* 36: 565–597.
200. Jones, M.P. 2009. Entrepreneurial Science: The Rules of the Game. *Social Studies of Science* 39: 821–851.
201. Schmidt, E. 2006. The Biotech Analyst's View. *Nature Biotechnology* 24: 261–262.
202. Farrell, M. 2008. Universities that Turn Research into Revenue. *Forbes*, September 12.
203. Perez-Pena, R. 2005. Research Scientist Gives$105 Million to N.Y.U. *New York Times*, August 12.
204. Wang, A.L. 2012. Northwestern University Leads Nation in Tech Transfer Revenue. *Crain's Chicago Business*, October 27.
205. Science Coalition. 2010. Sparking Economic Growth. *Science Coalition Report*, April.
206. Langreth, R. 2010. Why Biotech Investing is Like Gambling in Vegas. *Forbes*, December 21.
207. Glenn, M.A., and P.J. Nagle. 2010. Article I and the First Inventor to File: Patent Reform or Doublespeak? *IDEA- The Intellectual Property Law Review* 50: 441–462.
208. U.S. Constitution. art. I, cl. 8.
209. Abrams, D.S., and R.P. Wagner. 2013. Poisoning the Next Apple? The America Invents Act and Individual Inventors. *Stanford Law Review* 65: 517–563.
210. Murphy, A., M. Stramiello, J. Stroud, S. Lewis, and T. Irving. 2015. Impact of America Invents Act on Biotech Intellectual Property. *Cold Spring Harbor Perspectives in Medicine* 5.
211. Washburn, J. 2010. Big Oil Goes to College. *Center for American Progress*, October.
212. Wide, L. 2005. Inventions Leading to the Development of the Diagnostic Test Kit Industry – From the Modern Pregnancy Test to the Sandwich Assays. *Uppsala Journal of Medical Sciences* 110: 193–216.
213. Sturgeon, C.M., and E.J. McAllister. 1998. Analysis of hCG: Clinical Applications and Assay Requirements. *Annals of Clinical Biochemistry* 35: 460–491.
214. Cientifica. 2006. Where has My Money Gone? Government Nanotechnology Funding and the $18 Billion Pair of Pants. www.cientifica.com.
215. Lowenstein, H. 2001. Two Faces of State University Employment: Ethics in Access to Federal due Process. *Ethics & Behavior* 11: 39–53
216. Morrissey, S. 2003. Academic Immunity. *Chemical and Engineering News* 81: 42–43.
217. Office, G.A. 2010. College and University Endowments Have Shown Long-Term Growth, While Size, Restrictions, and Distributions. Vary GAO Report 10-393 February.
218. Hechinger, J. 2008. College Endowment Tax is Studied. *Wall Street Journal*, May 9.
219. Woodhouse, K. 2015. Who's to Blame for Rising Tuition? *Inside Higher Education*, May 5.
220. Belkin, D., and Thurm, S. 2012. Deans List: Hiring Spree Fattens College Bureaucracy – And Tuition. *Wall Street Journal*, December 28.
221. Greene, J.P., B. Kisida, and J. Mills. 2010. Administrative Bloat at American Universities: The Real Reason for High Costs in Higher Education. Goldwater Institute: Policy Report.
222. Johnson, A. 2009. Bowles Chides Chancellors for 'Embarrassing' Administrative Costs. *WRAL.com*, August 30.
223. Diethelm, P.A., J.C. Rielle, and M. McKee. 2005. The Whole Truth and Nothing but the Truth? The Research that Philip Morris did not Want you to see. *Lancet* 366: 86–92.

224. Genève, L.U.D. 2002. Affaire Rylander: l'Université de Genève tire ses conclusions. Université de Genève, Communiqué de Presse, December 20.
225. LaFrance, A. 2016. Scientists Brace for a Lost Generation in American Research. *Atlantic*, March 16.
226. Gannon, F. 2003. A Faustian Bargain for Universities? *Embo Reports* 4: 219–219.
227. Just, R.E., and W.E. Huffman. 2009. The Economics of Universities in a New Age of Funding Options. *Research Policy* 38: 1102–1116.
228. Buenstorf, G. 2009. Is Commercialization Good or Bad for Science? Individual-Level Evidence from the Max Planck Society. *Research Policy* 38: 281–292.
229. Glenna, L.L., R. Welsh, D. Ervin, W.B. Lacy, and D. Biscotti. 2011. Commercial Science, Scientists' Values, and University Biotechnology Research Agendas. *Research Policy* 40: 957–968.
230. Thursby, J.G., and M.C. Thursby. 2011. Has the Bayh-Dole Act Compromised Basic Research? *Research Policy* 40: 1077–1083.
231. Lowman, R.P. 2010. The Changing Role of Tenure at the American Research University. *Psychologist-Manager Journal* 13: 258–269.
232. Batterbury, S. 2008. Tenure or Permanent Contracts in North American Higher Education? A Critical Assessment. *Policy Futures in Education* 6: 286–297.
233. Koch, J. 2010. How Colleges Profit from High Textbook Prices. *New York Times*, July 29.
234. Brown, B.W. 2010. The Faux Value of E-Texts. *New York Times*, July 29.
235. Harkin, T.. 2010. Beware of for-Profit Higher Education. *Forbes*, August 11.
236. Wise, D., M.A. Stoto, K.S. Louis, M. Gluck, and D. Blumenthal. 1986. University-Industry Research Relationships in Biotechnology: Implications for the University. *Science* 232: 1361–1366.
237. Yassi, A., et al. 2010. The Good, the Bad, and the Ugly of Partnered Research: Revisiting the Sequestration Thesis and the Role of Universities in Promoting Social Justice. *International Journal of Health Services* 40: 485–505.
238. Boyd, E.A., and L.A. Bero. 2007. Defining Financial Conflicts and Managing Research Relationships: An Analysis of University Conflict of Interest Committee Decisions. *Science and Engineering Ethics* 13: 415–435.
239. Friedman, P.J. 2002. The Impact of Conflict of Interest on Trust in Science. *Science and Engineering Ethics* 8: 413–420.
240. Slaughter, S., S.L. Thomas, D.R. Johnson, and S.N. Barringer. 2014. Institutional Conflict of Interest: The Role of Interlocking Directorates in the Scientific Relationships Between Universities and the Corporate Sector. *Journal of Higher Education* 85: 1–35.
241. Hohm, C.F., and H.B. Shore. 1998. The Academy Under Siege: Informing the Public About the Merits of Academic Tenure. *Sociological Perspectives* 41: 827–831.
242. Sobel, D. 1995. *Longitude: The True Story of a Lone Genius Whole Solved the Greatest Scientific Problem of his Time*. New York: MacMillian.
243. Merton, R.K. 1968. The Matthew Effect in Science. *Science* 159: 56–63.
244. Gardner, S.K., and D. Veliz. 2014. Evincing the Ratchet: A Thematic Analysis of the Promotion and Tenure Guidelines at a Striving University. *Review of Higher Education* 38: 105–132.
245. Binder, R., A. Friedli, and E. Fuentes-Afflick. 2016. The New Academic Environment and Faculty Misconduct. *Academic Medicine* 91: 175–179.
246. Fanelli, D. 2010. Do Pressures to Publish Increase Scientists' Bias? An Empirical Support from US States Data. *PLos One* 5.
247. Montgomery, K., and A.L. Oliver. 2009. Shifts in Guidelines for Ethical Scientific Conduct: How Public and Private Organizations Create and Change Norms of Research Integrity. *Social Studies of Science* 39: 137–155.
248. Charlton, B.G. 2009. Are you an Honest Scientist? Truthfulness in Science Should be an Iron Law, not a Vague Aspiration. *Medical Hypotheses* 73: 633–635.

249. Lacetera, N., and L. Zirulia. 2012. The Economics of Scientific Misconduct (vol 27, pg 568, 2011). *Journal of Law Economics & Organization* 28: 183–183.
250. John, L.K., G. Loewenstein, and D. Prelec. 2012. Measuring the Prevalence of Questionable Research Practices with Incentives for Truth Telling. *Psychological Science* 23: 524–532.
251. Slaughter, S., and G. Rhoades. 1990. Renorming the Social Relations of Academic Science – Technology Transfer. *Educational Policy* 4: 341–361.
252. Evans, J.A. 2010. Industry Induces Academic Science to Know Less about More. *American Journal of Sociology* 116: 389–452.
253. Saiki, R., et al. 1988. Primer-Directed Enzymatic Amplification of DNA with a Thermostable DNA Polymerase. *Science* 239: 487–491.
254. Bower, D.J. 2003. Business Model Fashion and the Academic Spinout Firm. *R & D Management* 33: 97–106.
255. Gonzales, L.D., E. Martinez, and C. Ordu. 2014. Exploring Faculty Experiences in a Striving University Through the Lens of Academic Capitalism. *Studies in Higher Education* 39: 1097–1115.
256. Colombo, M.G., and E. Piva. 2008. Strengths and Weaknesses of Academic Startups: A Conceptual Model. *IEEE Transactions on Engineering Management* 55: 37–49.
257. Wall, M.E., and M.C. Wani. 1995. Camptothecin and Taxol – Discovery to Clinic – 13th Bruce F Cain Memorial Award Lecture. *Cancer Research* 55: 753–760.
258. Wall, M.E. and Wani, M.C. 1998. History and Future Prospects of Camptothecin and Taxol. In *Alkaloids*, ed. G.A. Cordell, vol. 50, 509–536. New York: Academic Press.
259. Faguet, G.B. 2006. *The War on Cancer: Anatomy of a Failure*. New York: Springer.
260. Kolata, G. 2013. Hopeful Glimmers in Long War on Cancer. *New York Times*, November 4.
261. Clark, M. 1977. Laetrile: Should It be Banned? *Newsweek* 89: 48.
262. Bitting, T.H. 1978. Drugs – Federal Drug Administration Ban on Laetrile Treatments for Terminally Ill Cancer Patients is Arbitrary and Capricious. *Tulsa Law Journal* 14: 222–225.
263. Moertel, C.G., et al. 1982. A Clinical Trial of Amygdalin (Laetrile) in the Treatment of Human Cancer. *The New England Journal of Medicine* 306: 201–206.
264. Howitz, K.T., et al. 2003. Small Molecule Activators of Sirtuins Extend Saccharomyces Cerevisiae Lifespan. *Nature* 425: 191–196.
265. Baur, J.A., et al. 2006. Resveratrol Improves Health and Survival of Mice on a High-Calorie Diet. *Nature* 444: 337–342.
266. Burnett, C., et al. 2011. Absence of Effects of Sir2 Overexpression on Lifespan in C. elegans and Drosophila. *Nature* 477: 482–485.
267. Couzin-Frankel, J. 2011. Aging Genes: The Sirtuin Story Unravels. *Science* 334: 1194–1198.
268. Doody, R.S., et al. 2008. Effect of Dimebon on Cognition, Activities of Daily Living, Behaviour, and Global Function in Patients with Mild-to-Moderate Alzheimer's Disease: A Randomised, Double-Blind, Placebo-Controlled Study. *Lancet* 372: 207–215.
269. Hopkins, C.R. 2010. ACS Chemical Neuroscience Molecule Spotlight on Dimebon. *Acs Chemical Neuroscience* 1: 587–588.
270. Reed, M.A., C. Zhou, C.J. Muller, T.P. Burgin, and J.M. Tour. 1997. Conductance of a Molecular Junction. *Science* 278: 252–254.
271. Overton, R. 2000. Molecular Electronics Will Change Everything. *Wired*, July.
272. Weiss, E.A., J.K. Kriebel, M.A. Rampi, and G.M. Whitesides. 2007. The Study of Charge Transport Through Organic Thin Films: Mechanism, Tools and Applications. *Philosophical Transactions of the Royal Society A-Mathematical Physical and Engineering Sciences* 365: 1509–1537.
273. Lieber, C.M. 2007. The Incredible Shrinking Circuit. *Scientific American* 17: 64–71.
274. Arkin, I.T. 2011. Science, Music, Literature and the One-Hit Wonder Connection. *Research Trends*, March.
275. Sampat, B.N. 2006. Patenting and US Academic Research in the 20th Century: The World Before and after Bayh-Dole. *Research Policy* 35: 772–789.

276. Carreyrou, J. 2015. Hot Startup Theranos has Struggled With Its Blood-Test Technology. *The Wall Street Journal*, October 16.
277. Kosloff, M. 2016. Theranos is Under Federal Criminal Investigation. *Vanity Fair*, April 19.
278. Tracy, A. 2016. Elizabeth Holmes's Theranos Nightmare Just Got Even Worse. *Vanity Fair*, May 26.
279. Greenspan, N. 2009. The Hype of Science: Leading Journals Including Science and Nature are Exaggerating Research Novelty. *The Scientist*, April 14.
280. Moult, J., J.T. Pedersen, R. Judson, and K.A. Fidelis. 1995. Large Scale Experiment to Assess Protein-Structure Prediction Methods. *Proteins-Structure Function and Genetics* 23: R2–R4.
281. Millikan, R.A. 1913. On the Elementary Electric Charge and the Avogadro Constant. *Physics Review* 2: 109–143.
282. Niaz, M. 2000. The Oil Drop Experiment: A Rational Reconstruction of the Millikan–Ehrenhaft Controversy and Its Implications for Chemistry Textbook. *Journal of Research in Science Teaching* 37: 480–505.
283. Fairbank, W.M., and A. Franklin. 1982. Did Millikan Observe Fractional Charges on Oil Drops. *American Journal of Physics* 50: 394–397.
284. Feynman, R.P., R. Leighton, and E. Hutchings. 1985. *Surely You' Re joking, Mr. Feynman*. New York: W. W. Norton.
285. Smith, G.D., and S. Ebrahim. 2002. Data Dredging, Bias, or Confounding – They Can All Get You into the BMJ and the Friday Papers. *British Medical Journal* 325: 1437–1438.
286. Selvin, H.C., and A. Stuart. 1966. Data-Dredging Procedures in Survey Analysis. *American Statistician* 20: 20–23.
287. Ioannidis, J.P.A. 2005. Why Most Published Research Findings are False. *PloS Medicine* 2: e124.
288. Rousseau, D.L. 1992. Case Studies in Pathological Science: How the Loss of Objectivity Led to False Conclusions in Studies of Polywater, Infinite Dilution and Cold Fusion. *American Scientist* 80: 54–63.
289. Derjaguin, B. 1983. Review of Polywater – Response. *Journal of Colloid and Interface Science* 94: 295–296.
290. Davenas, E., et al. 1988. Human Basophil De-granulation Triggered by Very Dilute Antiserum Again IgE. *Nature* 333: 816–818.
291. Wood, R.W. 1904. The n Rays. *Nature* 70: 530–531.
292. Yongsheng, L. 2004. Lysenko's Contributions to Biology and his Tragedies. *Rivista di Biologia/Biology Forum* 97: 483–498.
293. Gerber, J.S., and P.A. Offit. 2009. Vaccines and Autism: A Tale of Shifting Hypotheses. *Clinical Infectious Diseases* 48: 456–461.
294. Van Slyke, D.D., M. Vinograd-Villchur, and J.R. Losee. 1915. The Abderhalden Reaction. *Journal of Biological Chemistry* 23: 377–406.
295. Deichmann, U., and B. Müller-Hill. 1998. The Fraud of Abderhalden's Enzymes. *Nature* 393: 109–111.
296. Gratzer, W. 2008. The Golden Age of Fraud. *Biochemical Society*: 8–10.
297. Westfall, R.S. 1973. Newton and Fudge Factor. *Science* 179: 751–758.
298. Babbage, C. 1970. *Reflections on the Decline of Science in England*, 174–183. New York: Augustus M. Kelley.
299. Leonnig, C.D. 2010. CDC Misled District Residents About Lead Levels in Water, House Probe Finds. *Washington Post*, May 20.
300. Itkowitz, C. 2016. The Heroic Professor Who Helped Uncover the Flint Lead Water Crisis has been Asked to Fix it. *Washington Post*, January 27.
301. Fanelli, D. 2009. How Many Scientists Fabricate and Falsify Research? A Systematic Review and Meta-Analysis of Survey Data. *Plos One* 4.
302. Finetti, M., and A. Himmelrath. 1999. *The Fall of Man: Fraud and Falsification in German Science*. Ed. Raabe.
303. Naik, G. 2011. Mistakes in Scientific Studies Surge. *Wall Street Journal*, August 10.

304. Fox, M., and J. Beall. 2014. Advice for Plagiarism Whistleblowers. *Ethics & Behavior* 24: 341–349.
305. Schuklenk, U. 2012. Retraction Watch. *Bioethics* 26: ii.
306. Trikalinos, N.A., E. Evangelou, and J.P.A. Ioannidis. 2008. Falsified Papers in High-Impact Journals were Slow to Retract and Indistinguishable from Nonfraudulent Papers. *Journal of Clinical Epidemiology* 61: 464–470.
307. Nath, S.B., S.C. Marcus, and B.G. Druss. 2006. Retractions in the Research Literature: Misconduct or Mistakes? *Medical Journal of Australia* 185: 152–154.
308. Fang, F.C., R.G. Steen, and A. Casadevall. 2012. Misconduct Accounts for the Majority of Retracted Scientific Publications. *Proceedings of the National Academy of Sciences of the United States of America* 109: 17028–17033.
309. Habermann, B., M. Broome, E.R. Pryor, and K.W. Ziner. 2010. Research Coordinators' Experiences with Scientific Misconduct and Research Integrity. *Nursing Research* 59: 51–57.
310. Claxton, L.D. 2005. Scientific Authorship Part 1. A Window into Scientific Fraud? *Mutation Research-Reviews in Mutation Research* 589: 17–30.
311. Brimble, M., and P. Stevenson-Clarke. 2005. Perceptions of the Prevalence and Seriousness of Academic Dishonesty in Australian Universities. *Australian Educational Researcher* 32: 19–44.
312. Akst, J. 2010. When is Self-Plagiarism Ok? *New Scientist*, September 9.
313. Lewis, J., S. Ossowski, J. Hicks, M. Errami, and H.R. Garner. 2006. Text Similarity: An Alternative Way to Search MEDLINE. *Bioinformatics* 22: 2298–2304.
314. V.B. Institute. Deja Vu: A Database of Highly Similar Citations. http://spore.vbi.vt.edu/dejavu/.
315. Reich, E.R. 2010. Self-Plagiarism Case Prompts Calls for Agencies to Tighten Rules. *Nature* 568: 745.
316. Editor. 2013. Retraction Blues. *Nature Medicine* 19: 1547–1548.
317. Errington, T.M., et al. 2014. An Open Investigation of the Reproducibility of Cancer Biology Research. *Elife* 3: e04333.
318. Fanelli, D. 2018. Is Science Really Facing a Reproducibility Crisis, and Do We Need It To? *Proceedings of the National Academy of Sciences of the United States of America* 115: 2628–2631.
319. Mullard, A. 2011. Reliability of 'New Drug Target' Claims Called into Question. *Nature Reviews Drug Discovery* 10: 643–644.
320. Huston, L. 2013. Lancet Formally Retracts Jikei Heart Study of Valsartan. *Forbes*, September 5.
321. Naik, G. 2011. Scientists' Elusive Goal: Reproducing Study Results. *Wall Street Journal*, December 2.
322. Begley, S. 2012. In Cancer Science, Many "Discoveries" Don't Hold Up. *Reuters Edition*, March 28.
323. Alexander, A., et al. 2012. An Open, Large-Scale, Collaborative Effort to Estimate the Reproducibility of Psychological Science. *Perspectives on Psychological Science* 7: 657–660.
324. Aarts, A.A., et al. 2015. Estimating the Reproducibility of Psychological Science. *Science* 349: aac4716.
325. Button, K.S., et al. 2013. Power Failure: Why Small Sample Size Undermines the Reliability of Neuroscience. *Nature Reviews Neuroscience* 14: 365–376.
326. Carlisle, J.B. 2017. Data Fabrication and Other Reasons for Non-random Sampling in 5087 Randomised, Controlled Trials in Anaesthetic and General Medical Journals. *Anesthesia* 72: 944–952.
327. MacDougall, M. 2014. Assessing the Integrity of Clinical Data: When is Statistical Evidence Too Good to be True? *Topoi-an International Review of Philosophy* 33: 323–337.
328. Keller, M.B., et al. 2001. Efficacy of Paroxetine in the Treatment of Adolescent Major Depression: A Randomized, Controlled Trial. *Journal of the American Academy of Child and Adolescent Psychiatry* 40: 762–772.

329. Jureidini, J.N., and L.B. McHenry. 2011. Conflicted Medical Journals and the Failure of Trust. *Accountability in Research-Policies and Quality Assurance* 18: 45–54.
330. DeAngelis, C.D. 2006. The Influence of Money on Medical Science. *JAMA* 296: 996–998.
331. Horvitz, L.A. 1994. Can Scientists Police Themselves? *Business Library*, May 2.
332. Dingell, J.D. 1996. The Elusive Truths of the Baltimore Case. *Washington Post*, July 18, A27.
333. Pain, E. 2014. Paul Brookes: Surviving as an Outed Whistleblower. *Science*, December 10.
334. Frezza, B. 2013. A Barrage of Legal Threats Shuts Down Whistleblower Site, Science Fraud. *Forbes*, January 9.
335. da Silva, J.A.T., A. Al-Khatib, and J. Dobranszki. 2017. Fortifying the Corrective Nature of Post-Publication Peer Review: Identifying Weaknesses, Use of Journal Clubs, and Rewarding Conscientious Behavior. *Science and Engineering Ethics* 23: 1213–1226.
336. Servick, K. 2014. Researcher Files Lawsuit over Anonymous PubPeer Comments. *ScienceInsider*, October 26.
337. Flaherty, C. 2014. PubPeer Pressure. *Inside Higher Education*, November 3.
338. Dyer, O. 2016. Researcher Suing PubPeer was Found Culpable of Misconduct, Court Documents Show. *BMJ* 355: i6337.
339. Buranyi, S. 2016. Anonymous Internet Vigilantes are Taking Peer Review into Their Own Hands. *Motherboard*, October 3.
340. Shea, C. 2012. The Data Vigilante. *The Atlantic*, December.
341. Oransky, I. 2013. McGill Committee Says Nature Figures were "Intentionally Contrived and Falsified". *Retraction Watch*, January 25.
342. Suzuki, M., et al. 2006. RETRACTED: Methylation of Apoptosis Related Genes in the Pathogenesis and Prognosis of Prostate Cancer. *Cancer Letters* 242: 222–230.
343. Scudellari, M. 2015. ORI-Sanctioned Former UT-Southwestern Cancer Researchers Up to 10 Retractions. *Retraction Watch*, May 18.
344. Toth, C., et al. 2008. RETRACTED: Local Erythropoietin Signaling Enhances Regeneration in Peripheral Axons. *Neuroscience* 154: 767–783.
345. Oransky, I. 2014. Diabetes Researcher Cory Toth now up to Nine Retractions. *Retraction Watch*, September 2.
346. Sprague, R.L. 1993. Whistleblowing: A Very Unpleasant Avocation. *Ethics & Behavior* 3: 105–133.
347. Oransky, I. 2013. Emory Cardiology Researcher Up to Six Retractions. *Retraction Watch*, December 26.
348. Ushio-Fukai, M., et al. 2001. Cholesterol Depletion Inhibits Epidermal Growth Factor Receptor Transactivation by Angiotensin II in Vascular Smooth Muscle Cells. *Journal of Biological Chemistry* 276: 48269–48275.
349. Marcus, A. 2013. Another Image Falsification Retraction for Emory Heart Researchers. *Retraction Watch*, November 12.
350. Ushio-Fukai, M., et al. 2005. CAbl tyrosine kinase Mediates Reactive Oxygen Species and Caveolin-Dependent AT(1) Receptor Signaling in Vascular Smooth Muscle – Role in Vascular Hypertrophy (Retracted article. See vol. 106, pg. 1784, 2010). *Circulation Research* 97: 829–836.
351. Zuo, L., M. Ushio-Fukai, N.A. Patrushev, and R.W. Alexander. 2004. Caveolin-1 is Required for AT(1) Receptor Trafficking into Caveolae/Lipid Rafts Coupled to Redox Signaling in Vascular Smooth Muscle Cells. *Circulation* 110: 89–89.
352. Zuo, L., M. Ushio-Fukai, L.L. Hilenski, and R.W. Alexander. 2004. Microtubules Regulate Angiotensin II Type 1 Receptor and Rac1 localization in Caveolae/Lipid Rafts – Role in Redox Signaling (Retracted article. See vol. 31, pg. E10, 2011). *Arteriosclerosis Thrombosis and Vascular Biology* 24: 1223–1228.
353. Ushio-Fukai, M., et al. 2002. Novel Role of gp91(phox)-Containing NAD(P)H Oxidase in Vascular Endothelial Growth Factor-Induced Signaling and Angiogenesis. *Circulation Research* 91: 1160–1167.

354. Ushio-Fukai, M. et al. 2001. Epidermal Growth Factor Receptor Transactivation by Angiotensin II Requires Reactive Oxygen Species in Vascular Smooth Muscle Cells. *Arteriosclerosis Thrombosis and Vascular Biology* 21: 489–495.
355. Anumolu, R., B.J. Robinson, and L.F. Pease. 2013. Chopstick Nanorods: Tuning the Angle Between Pairs with High Yield. *Nano Letters* ASAP (retracted). https://doi.org/10.1021/nl400959z.
356. Anumolu, R., et al. 2011. Fabrication of Highly Uniform Nanoparticles from Recombinant Silk-Elastin-like Protein Polymers for Therapeutic Agent Delivery. *ACSNano* 5: nn53103585f.
357. Hayden, E.C. 2008. Chemistry: Designer Debacle. *Nature News* 453: 275–278.
358. Schulz, W.G. 2011. A Puzzle Named Bengü Sezen. *Chemical and Engineering News* 89: 40–43.
359. Schultz, W.G. 2011. Reports Detail a Massive Case of Fraud. *Chemical and Engineering News*, July 11.
360. Lin, S.Q. 2013. Why Serious Academic Fraud Occurs in China. *Learned Publishing* 26: 24–27.
361. Zeng, W.Q., and D. Resnik. 2010. Research Integrity in China: Problems and Prospects. *Developing World Bioethics* 10: 164–171.
362. Stafford, N. 2010. China: Act on Scientific Fraud. *Royal Society of Chemistry*, January 15.
363. National Editorial Board. 2010. A Hammer Blow to National Ethics. *Nature* 467: 884.
364. Martinson, B.C., M.S. Anderson, and R. de Vries. 2005. Scientists Behaving Badly. *Nature* 435: 737–738.
365. Nilstun, T., R. Lofmark, and A. Lundqvist. 2010. Scientific Dishonesty-Questionnaire to Doctoral Students in Sweden. *Journal of Medical Ethics* 36: 315–318.
366. Cyranoski, D. 2012. Research Ethics: Zero Tolerance a University Cracks Down on Misconduct in China. *Nature News*, January 11.
367. Abbott, A. 1998. German Oncology Research Shaken by Fraud Case. *Annals of Oncology* 9: 1–2.
368. Hall, C. 1994. Consultant Sacked Over Pregnancy Claims. *The Independent*, December 24.
369. Reich, E. 2011. Fresh Dispute about MMR 'Fraud'. *Nature* 479: 157–115.
370. Whalen, J., and B. McKay. 2013. Fifteen Years After Autism Panic, a Plague of Measles Erupts. *Wall Street Journal*, July 19.
371. Deer, B. 2006. MMR Doctor Given Legal Aid Thousands. *The Times (London)*, December 31.
372. Fleischmann, M., and S. Pons. 1989. Electrochemically Induced Nuclear Fusion of deuterium. *Journal of Electroanalytical Chemistry and Interfacial Electrochemistry* 261: 301–308.
373. Reich, E.S. 2007. Congress Requests Bubble-Fusion Reports. *Nature* 446: 480–480.
374. Reich, E.S. 2008. Fusion Verdict: Misconduct. *Nature* 454: 379–379.
375. Bonnefoi, H., et al. 2007. Validation of Gene Signatures that Predict the Response of Breast Cancer to Neoadjuvant Chemotherapy. *The Lancet Oncology* 8: 1071–1078.
376. Economist Editorial. 2011. An Array of Errors. *The Economist*, September 10.
377. Rupp, B. 2012. Detection and Analysis of Unusual Features in the Structural Model and Structure-Factor Data of a Birch Pollen Allergen. *Acta Crystallographica Section F-Structural Biology Communications* 68: 366–376.
378. Stanfield, R., E. Pozharski, and B. Rupp. 2016. Additional Comment on Three X-ray Crystal Structure Papers. *Journal of Immunology* 196: 528–530.
379. Mishkin, B. 1999. Scientific Misconduct: Present Problems and Future Trends. *Science and Engineering Ethics* 5: 283–292.
380. Oransky, I. 2014. Science hasn't Retracted Paper that University, NSF Investigators Wanted Withdrawn. *Retraction Watch*, January 30.
381. McCook, A. 2016. After 10 Years, a Whistleblower is Vindicated. Here's Why He Kept Going. *Retraction Watch*, February 10.
382. FRANZEN vs. North Carolina State University Petition for Public Records in Wake County Superior Court. 2016. Case No. 16-CVS-12363.

383. Carter, C.J., et al. 2009. In vitro Selection of RNA Sequences Capable of Mediating the Formation of Iron Oxide Nanoparticles. *Journal of Materials Chemistry* 19: 8320–8326.
384. Han, A.P. 2017. ORI: Ex-grad Student "Falsified and/or Fabricated" Data in PNAS Submission. *Retraction Watch*, December 11.
385. Palus, S. 2015. 4th ORI-Flagged Paper by Oregon Student is Retracted. *Retraction Watch*, September 28.
386. Stern, V. 2017. Another Retraction for Student Who Confessed to Cooking Data. *Retraction Watch*, April 13.
387. Braxton, J.M., and L.L. Baird. 2001. Preparation for Professional Self-Regulation. *Science and Engineering Ethics* 7: 593–610.
388. Allison, D.B., A.W. Brown, B.J. George, and K.A. Kaiser. 2016. A Tragedy of Errors. *Nature* 530: 27–29.
389. Reich, E. 2006. Bad Data Fail to Halt Patents. *Nature News*, January 26.
390. Weaver, D., et al. 1986. Altered Repertoire of Endogenous Immunoglobulin Gene Expression in Transgenic Mice Containing a Rearranged Mu Heavy Chain Gene. *Cell* 45: 247–259.
391. Resnik, D.B. 2003. From Baltimore to Bell Labs: Reflections on Two Decades of Debate about Scientific Misconduct. *Accountability in Research* 10: 123–135.
392. Hilts, P.J. 1992. The Science Mob: The David Baltimore Case-and its Lessons. *The New Republic* 25: 28–31.
393. Fish, A.L. 1991. NIH Report Finds Fraud in MIT Research. *The Tech* 111, March 22.
394. Travis, J. 1993. Imanishi-Kari Says Her New Data Shows She is Right. *Science* 260: 1073–1074.
395. Opinion. 1996. The Fraud Case that Evaporated. *New York Times*, June 25.
396. Stone, R., and E. Marshall. 1994. Imanishi-Kari Case: ORI Finds Fraud. *Science* 266: 1468–1469.
397. Lang, S. 1993. Questions of Scientific Responsibility: The Baltimore Case. *Ethics & Behavior* 3: 3–72.
398. Hotz, R.L. 2005. Caltech President Who Raised School's Profile to Step Down. *Los Angeles Times*, October 4, A1.
399. Reich, E. 2011. Fraud Case We Might Have Seen Coming. *Nature News*, July 28.
400. Reich, E. 2011. Biologist Spared Jail for Grant Fraud. *Nature News*, June 28.
401. Price, A.R. 2013. Research Misconduct and Its Federal Regulation: The Origin and History of the Office of Research Integrity-with Personal Views by ORI's Former Associate Director for Investigative Oversight. *Accountability in Research-Policies and Quality Assurance* 20: 291–319.
402. U.S. Government. 2005. 42 CFR Parts 50 and 93. *Federal Register* 70: 28369–28400.
403. Government. 2005. U.S. 10 CFR Parts 600 and 733. *Federal Register* 70: 37010–37016.
404. Reich, E. 2011. Misconduct Oversight at the DOE: Investigation Closed. *Nature* 475: 20–22.
405. Bonito, A.J., et al. 2012. Preparing Whistleblowers for Reporting Research Misconduct. *Accountability in Research-Policies and Quality Assurance* 19: 308–328.
406. Breen, K.J. 2016. Research Misconduct: Time for a Re-think? *Internal Medicine Journal* 46: 728–733.
407. Reich, E.S. 2009. Bubble-Fusion Scientist Debarred from Federal Funding. *Nature News*.
408. Odling-Smee, L., J. Giles, I. Fuyuno, D. Cyranoski, and E. Marris. 2007. Misconduct Special: Where are They Now? *Nature* 445: 244–245.
409. Scudellari, M. 2015. https://retractionwatch.com/category/by-author/gazdar/. *Retraction Watch*.
410. Petsko, G.A. 2007. And the Second Shall Be First. *Genome Biology* 8: 103.
411. Caldeira, K. 2016. RL and the Impact of Small Groups of Authors. *Environmental Research Letters* 11: 120204.
412. North Carolina General Statutes § 132-6 and § 126-84, et.seq.
413. McCluskey, M. 2017. Public Universities Get an Education in Private Industry. *The Atlantic*, April 3.

414. Krimsky, S., L.S. Rothenberg, P. Stott, and G. Kyle. 1998. Scientific Journals and Their Authors' Financial Interests: A Pilot Study. *Psychotherapy and Psychosomatics* 67: 194–201.
415. Bekelman, J.E., Y. Li, and C.P. Gross. 2003. Scope and Impact of Financial Conflicts of Interest in Biomedical Research: A Systematic Review. *JAMA* 289: 454–465.
416. Slaughter, S., M.P. Feldman, and S.L. Thomas. 2009. US Research Universities' Institutional Conflict of Interest Policies. *Journal of Empirical Research on Human Research Ethics* 4: 3–20.
417. Gunsalus, C.K. 2019. Make Reports of Research Misconduct Public. *Nature* 570: 7.
418. Gunsalus, C.K., A.R. Marcus, and I. Oransky. 2018. Institutional Research Misconduct Reports Need More Credibility. *JAMA* 319: 1315–1316.
419. Olivieri, N.F. 2003. Patients' Health or Company Profits? The Commercialisation of Academic Research. *Science and Engineering Ethics* 9: 29–41.
420. Basken, P. 2009. Science Ethics Rules Leave Room for Scandals, Critics Fear. *The Chronicle of. Higher Education*, July 20.
421. Friedman, P.J. 1990. Correcting the Literature Following Fraudulent Publication. *JAMA* 263: 1416–1419.
422. Peterson, E. 2011. Court Reinstates Professor's Libel Suit Against Ohio University. *The Reporters Committee for Freedom of the Press*, July 27.
423. DeMets, D.L., T.R. Fleming, G. Geller, and D.F. Ransohoff. 2017. Institutional Responsibility and the Flawed Genomic Biomarkers at Duke University: A Missed Opportunity for Transparency and Accountability. *Science and Engineering Ethics* 23: 1199–1205.
424. Broad, M.C. 2008. Endowments Are both Vital and Misunderstood. *The Chronicle of Higher Education*, November 21.
425. Symonds, W.C. 2004. Rich College, Poor College. *Bloomberg Business Week*, December 20.
426. Madey. Madey v. Duke University. 2002. *Court of Appeals, Federal Circuit* 307 F.3d 1351.
427. Miller, J. 2003. Sealing the Coffin on the Experimental Use Exception. *Duke Law and Technology Review (Duke School of Law, Duke University)* 2.
428. Wilson, D. 2009. Harvard University in Ethics Quandary. *New York Times*, March 2.
429. Herson, J. 2016. Strategies for Dealing with Fraud in Clinical Trials. *International Journal of Clinical Oncology* 21: 22–27.
430. Woogan, D. 2012. Industry Money and Questionable Ethics Contaminate UT Austin Fracking Study. *Scientific American* July 24.
431. Pachter, W.S., R.E. Fox, P. Zimbardo, and D.O. Antonuccio. 2007. Corporate Funding and Conflicts of Interest – A Primer for Psychologists. *American Psychologist* 62: 1005–1015.
432. Dalton, R. 2008. Modified Genes Spread to Local maize. *Nature* 456: 149.
433. Milius, S. 2001. Journal Disowns Transgene Report. *Science News Online*, November 26.
434. Dalton, R. 2004. Review of Tenure Refusal Uncovers Conflicts of Interest. *Nature* 430: 598.
435. Hakim, D. 2016. Scientists Loved and Loathed by an Agrochemical Giant. *New York Times*, December 31.
436. Report. 2012. Public Research, Private Gain. *Food and Water Watch*, April.
437. Kahr, B., and M.D. Hollingsworth. 2019. Massive Faculty Donations and Institutional Conflicts of Interest. *The Journal of Scientific Practice and Integrity* 1: 740579.
438. Goldstein-Street, J. 2019. Internal Emails Unearthed in New Report Shed Light on Claims of Scientific Misconduct Among UW Chemists. *The Daily of the University of Washington*, May 22, http://www.dailyuw.com/news/article_17fa7902-7c2e-11e9-949d-53b97af457d8.html
439. B. Kahr, M. D. Hollingsworth Misconduct investigators, show your work Chemistry World, July 31, 2019. [https://www.chemistryworld.com/opinion/misconduct-investigators-show-your-work/3010725.article]
440. Mullins, B., D. Belkin, and A. Fuller 2015. Colleges Flex Lobbying Muscle. *Wall Street Journal*, November 8.
441. Cohen, R. 2014. Universities Pay Plenty for Influence and Access Through Lobbying. *Nonprofit Quarterly*, July 16.

442. Gottschalk, P. 2017. Private Police Legitimacy: The Case of Internal Investigations by Fraud Examiners. *Policing-an International Journal of Police Strategies & Management* 40: 628–640.
443. National Academies of Science, Engineering, and Medicine. 2017. *Fostering Integrity in Research.* Washington D.C.: The National Academies Press, 326p. http://sites.nationalacademies.org/PGA/PGA178432.
444. Crocker, J. 2011. The Road to Fraud Starts with a Single Step. *Nature* 479: 151.
445. Hasselmo, N. 2002. Individual and Institutional Conflict of Interest: Policy Review by Research Universities in the United States. *Science and Engineering Ethics* 8: 421–427.
446. Kochan, C.A., and J.M. Budd. 1992. The Persistence of Fraud in the Literature – The Darsee Case. *Journal of the American Society for Information Science* 43: 488–493.